MENSA SCIENCE EXPERIMENTS

Bob Bonnet & Dan Keen

**OFFICIAL MENSA
PUZZLE BOOK**

Main Street
A division of Sterling Publishing Co., Inc.
New York

Library of Congress Cataloging-in-Publication Data Available

2   4   6   8   10   9   7   5   3   1

This collection is excerpted from the following Sterling titles:
Science Fair Projects Chemistry © 2000
by Bob Bonnet and Dan Keen
Science Fair Projects Energy © 1997
by Bob Bonnet and Dan Keen
Science Fair Projects Physics © 1999
by Bob Bonnet and Dan Keen

Published by Sterling Publishing Co., Inc.
387 Park Avenue South, New York, NY 10016
© 2004 by Sterling Publishing Co., Inc.
Distributed in Canada by Sterling Publishing
c/o Canadian Manda Group, One Atlantic Avenue, Suite 105
Toronto, Ontario, Canada M6K 3E7
Distributed in Great Britain and Europe by Chris Lloyd at Orca Book
Services, Stanley House, Fleets Lane, Poole BH15 3AJ, England
Distributed in Australia by Capricorn Link (Australia) Pty. Ltd.
P.O. Box 704, Windsor, NSW 2756, Australia

ISBN 1-4027-1642-7

CONTENTS

Introduction to Science Experiments

Because safety is and must always be the first consideration, we recommend that ALL activities be done under adult supervision. Even seemingly harmless objects can become a hazard under certain circumstances. If you can't do a project safely, then don't do it!

Respect for life should be fundamental. Your project cannot be inhumane to animals. Disruption of natural processes should not occur thoughtlessly and un-necessarily. Interference with ecological systems should always be avoided. Ethical rules must be followed also. It is unethical to hypothesize that one race or religion is better than another.

Science is the process of finding out. "The scientific method" is a procedure used by scientists and students in science fairs. It consists of several steps: identifying a problem or stating a purpose, forming a hypothesis, setting up an experiment to collect information, recording the results and coming to a conclusion as to whether or not the stated hypothesis is correct.

A science project starts by identifying a problem, asking a question or stating a purpose. The statement of the problem defines the boundaries of the investigation. For example, air pollution is a problem, but you must set the limits of your project. It is unlikely that you have access to an electron microscope, so an air pollution project could not check pollen in the air. This project might be limited to the accumulation of dust and other visible materials.

Once the problem is defined, a hypothesis (an educated guess about the results) must be formed. Hypothesize that there is more dust in a room that has thick carpeting than in a room that has hardwood or linoleum flooring.

Often, a hypothesis can be stated in more than one way. For example, in considering a project to gather data for using rocks to store and release heat during the night in a solar heated home, you might test to see if one large rock or many smaller rocks are better for giving off stored heat for a longer period of time. This could be stated in two ways: Hypothesize that one large rock will give off stored heat for a longer period of time than an equal mass of smaller rocks. Or, you could state the opposite: Hypothesize that smaller rocks will give off stored heat for a longer period of time compared to one large rock of equal mass. It does not matter which way the hypothesis is stated, nor does it matter which one is correct. The hypothesis doesn't have to be proven correct in order for the project to be a success; it is successful if facts are gathered and knowledge is gained.

Then you must set up an experiment to test your hypothesis. You will need to list materials, define the variables, constants and any assumptions, and document your procedure so that you or someone else will be able to repeat the experiment at another time. Finally, from the results collected, you must come to a conclusion as to whether or not the hypothesis is correct.

When choosing a science fair topic, pick something that is interesting to you, that you would like to work on. Then all of your research and study time will be spent on a subject you enjoy!

For presentation at a formal science fair, consider early on how you can demonstrate your project. Remember, you may not be able to control certain conditions in a gym or a hall. Decide how to display the project's steps and outcome, and keep a log or journal of how you got your results and came to your conclusion (photographs or even a video). Something hands-on or interactive often adds interest to a project display.

Bob Bonnet & Dan Keen

CHEMISTRY

CHEMISTRY

Chemistry

Welcome to the world of chemistry! This section explores projects in this fascinating field. Chemistry is the study of what substances are made of, how they can be changed and how they can be combined with other substances to make new substances. When substances are changed or combined, their old properties are changed and new ones are taken on. Plastic is an example of a new substance made from other substances.

Chemistry is a branch of physical science. It is one of the most interesting and motivating topics in science. It's important to understand chemistry because so many of its principles are found in other science disciplines, including astronomy, geology, mathematics, environmental science, botany, health and medicine, electronics, physics and even the arts. Some of its basic principles are safety, measurement, the scientific method of procedures and evaluation, cause and effect, written reporting of results and problem recognition.

Chemistry is at work all around us, and is a part of our daily lives. Medicines, garden fertilizers, food preservatives, synthetics, energy from batteries and the burning of coal to make electricity, glass made from limestone and sand, explosives, disinfectants for cleaning, and even the baking of yeast to cause bread to rise are all examples of chemistry at work. Read the label on packaged foods and cleaning supplies and you will often see listed the names of chemicals that are contained in it.

Energy can be given off from chemical changes: electricity, light and heat. When something burns, a chemical reaction is taking place, with light and heat given off. Chemical changes can occur when elements come in contact with each other, when they decompose or when temperature or pressure are changed.

UNDER PRESSURE
How handling affects substances

Purpose To study the effects of squeezing on liquid, a solid, some gel and a colloidal substance.

Overview Pressure sometimes causes a change in matter. Great underground pressure is what creates diamonds. But even squeezing by hand can cause change in certain substances. Cornstarch, made from corn and used in cooking as a thickening agent (as in gravy), is what is called a colloidal substance. It is made up of small particles that don't dissolve but stay suspended in a fluid. Mixed with water, when cornstarch is at rest, it forms a substance that somewhat resembles a liquid, but the substance

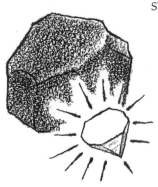

changes its property to be more like a solid when pressure is applied.

Hypothesis Of four sample materials, only the colloidal substance will undergo physical change (other than breaking up) when hand pressure is applied.

Procedure Take some gelatin dessert from the refrigerator, place it into the palm of your hand and use the forefinger and thumb of your other hand to apply pressure. The CONSTANT is the application of this pinching pressure. The VARIABLES are the materials being tested. Now, apply pressure to an ice chip or cube, and to some water.

Next, shake about a teaspoon of cornstarch into your palm. Add a few drops of water to the cornstarch, and stir it around to mix it in. Slowly add a few more drops, a little at a time. When the mixture is slightly watery, it's ready. Now, squeeze the colloidal substance between your forefinger and thumb and the pressure causes it to become putty-like. Stop, and it becomes more liquid, with a little water seeping into your palm. Do it again. Doesn't it have a strange feel? What a great natural toy!

Results & Conclusion   Write down the results of your experiment. Come to a conclusion as to whether or not your hypothesis was correct.

Something more   Does a flour-and-water mixture become a colloidal substance?

NOTHING TO SNIFF AT
Smell and the secret of "wafting"

Purpose How to smell a strong substance safely.

Overview Smells come from in-air particles in the form of gases and vapors. The chemical solution and nerve cells that line our noses interact with the

vapors, and the brain interprets the smell.

Although some scents, like flowers, cinnamon, and coffee, are nice, others are unpleasant and can even be hazardous. Breathing gas, for example, given off by powdered chlorine granules used in swimming pools when even a little water is spilled on them, can make you deathly ill. *Never* inhale, or take a deep breath, of any substance that you are not perfectly sure is completely safe!

A safety practice in chemistry of just "sampling" a scent, which should be used throughout your life, is called wafting. It's a way to get just a "whiff" of the odor a substance gives off. A very small amount is mixed, with a wave of your hand, into a puff of air—then you quickly "sniff."

Hypothesis You can "sample" an odor in a way that avoids physical discomfort.

Procedure This science project demonstrates the seriousness, and danger, of breathing in substances. The CONSTANT is the human olfactory system—the nose. The VARIABLE is the method used to introduce the onion odor to your nose.

First, have an adult chop up a strong fresh onion and close it in a plastic bag. Open the bag near your face. Quickly wave your hand over the opening, fanning some of the onion vapors past your nose. Sniff, and close the bag. Did you smell the onion? Now open the bag and take in a good deep breath. You'll probably cough and your eyes will tear up. Which do you think is the safer way to smell the vapor?

Results & Conclusion Write down the results of your experiment. Come to a conclusion as to whether or not your hypothesis was correct.

Something more Try wafting other unpleasant-smelling, *safe* (food) substances.

ACUTE CUKE
The chemical change called pickling

Purpose Some like sweet pickles, some like sour ones, some prefer regular cucumbers. What's the difference in their pH readings?

YOU NEED

- dill pickle
- fresh cucumber
- litmus paper
- color comparison chart for litmus paper
- 2 spoons

Overview Pickling is a chemical process used to preserve food. Fruits, vegetables and even meats can be pickled. The food to be preserved is soaked in vinegar and brine (brine is very salty water). Sometimes sugar and spices are added to change the flavor.

The most common food that is processed that way is the cucumber, which is then called a "pickle." Dill pickles are the most common type of sour-tasting pickles. Small pickled cucumbers, called gherkins, are known for their sweet taste.

PH (potential of electricity for positive hydrogen ions) is a measure of a substance's acidity. On the pH scale, pure water is in the middle, with a pH of 7. A lower number means the substance is an acid; a higher number means it is an alkali. Litmus paper is a device which turns red when dipped in a substance that is an acid and blue when it is an alkali. Does a food that tastes sour have a low pH, meaning it is an acid?

Hypothesis Hypothesize that a sour-tasting pickle will have a low pH.

Procedure Squeeze some cucumber juice onto a spoon and some pickle juice onto another spoon. Use a strip of litmus paper to test the pH of the cucumber juice, then the juice of the dill pickle. Which is more acidic? Which tastes more sour? The CONSTANTS are your taste buds and the litmus paper pH indicators. The VARIABLE is the state of the cucumber—fresh or dilled.

Results & Conclusion   Write down the results of your experiment. Come to a conclusion as to whether or not your hypothesis was correct.

Something more   Do sweeter pickles, such as gherkins, have low pH? Test several kinds of pickle after hypothesizing which will have lower pH. Then eat them.

GETTING AHEAD

Reaction between a base and an acid

Purpose To learn if lemon juice can replace vinegar in the traditional "volcano eruption" project?

Overview When baking soda and vinegar come together, a chemical reaction takes place. (Baking soda is a "base" and vinegar is an "acid.") Carbon dioxide gas, known as CO_2, is quickly released. Bubbles foam up and spill out of the container in a violent eruption that's impressive but an all-too-common project. But, is there something special about vinegar that causes this reaction, or is it simply the fact that vinegar is an acid?

Hypothesis Because vinegar and lemon juice are both acids, their reactions to baking soda will be similar.

YOU NEED
- baking soda
- lemon juice
- vinegar
- 2 same-size drinking glasses
- teaspoon
- a sink

Procedure Pour some vinegar into one glass and the same amount of lemon juice into the other. Holding the glass with vinegar over a sink, add a teaspoonful of baking soda. A chemical reaction takes place, releasing CO_2—like opening a soda can that has been shaken. Notice the size of the bubble "head."

Then, hold the glass with the lemon juice—an acid, too—over the sink. Add the teaspoonful of baking soda. The baking soda is CONSTANT. The VARIABLE is the type of acids used, vinegar and lemon juice. Does the lemon juice solution bubble?

Results & Conclusion Write down the results of your experiment. Come to a conclusion as to whether or not your hypothesis was correct.

Something more

1. Try other substances with acid pH: orange, grapefruit or other citrus juices. Check their pHs first. Is the reaction bigger with lower pH juices (more/less acidic)?

2. Does temperature affect the reaction of acid and baking soda? Put some pH acid in the refrigerator and an equal amount in a warm, sunny window. Once temperatures have adjusted, add an equal amount of baking soda and observe the reactions.

WE SALUTE SOLUTION
Understanding basic chemical terms

Purpose "Solutes" and "solvents" can chemically combine, other substances don't. True?

YOU NEED
- salt
- pepper
- water
- 2 drinking glasses
- teaspoon

Overview A "mixture" is two or more substances mixed together which do not chemically combine but remain the same as before. In a solution, a substance (called a solute) dissolves, becoming evenly distributed throughout another substance (called a solvent).

Hypothesis Hypothesize that pepper, which is not a solute, does not chemically combine with water. Salt, however, will combine with water.

Procedure Add a teaspoon of salt to a drinking glass filled with water. Stir. The salt dissolves into the water and can no longer be seen nor easily removed. The water and salt have chemically combined to form a "solution."

Add a teaspoon of pepper to another glass filled with water. Stir. Has the pepper dissolved in the water, or is it floating on top or rising slowly within it? Has the pepper stayed essentially the way it was before it was added to the water? Pepper and water make a "mixture." The water and the amount of material added to the water are CONSTANT. The type of material added to the water is varied (salt, then pepper).

Results & Conclusion   Write down the results of your experiment. Come to a conclusion as to whether or not your hypothesis was correct.

Something more

1. Can you add salt to a glass of water until so much is added that no more will dissolve into the water, and the salt instead settles to the bottom of the glass?

2. You can separate pepper from water by skimming it off the surface. Is it possible to recover salt from water? What if the water were evaporated?

3. Adding powdered chocolate to warm milk to make hot cocoa is a solution. Does adding sugar to iced tea make a solution or a mixture?

GO, OLD MOLD!

Food additives keep bread good longer

Purpose To determine which brands and types of bread you buy have chemical mold inhibitors added to them.

Overview A science experiment traditionally done in elementary school grades involves growing mold on a slice of moist bread. Molds are microscopic plants. Molds grow from tiny particles called spores, which travel through the air. A moist slice of bread is an excellent "home" for mold to grow.

YOU NEED

- packaged white bread
- fresh white bread (home-baked or local bakery)
- rye bread
- wheat bread
- water
- four plates
- teaspoon
- pencil
- paper

However, in recent years teachers are finding that growing mold on bread isn't always easy! The reason is that many breads today have some kind of special chemical added to them to stop mold. It is called a mold inhibitor.

Food additives are substances added to foods during processing to either help preserve them, improve color or flavor, or make their texture more appealing. Chemists have also devised food additives that inhibit (slow down) the growth of molds, and some of those additives are commonly placed in packaged bread. These food additives have passed many tests and have been approved by the United States Food and Drug Administration as being safe to eat before manufacturers were permitted to add them to their food products.

By adding such mold-inhibitor chemicals to breads, today's baked loaves will not go moldy and will remain edible for a longer period of time.

Does fresh-baked bread from a local bakery provide a better medium (place) for growing mold than mass-processed, packaged bread from a supermarket, which may contain food additives as preservatives?

Hypothesis Hypothesize that some breads are made with mold inhibitors added, so mold will not grow on them as quickly as on other breads.

Procedure Select four slices of bread and place each on a plate. Each slice should be a different kind of bread. One slice should be from a fresh-baked loaf from a local bakery. One should be a packaged white bread, another a slice of rye bread and another a slice of wheat bread. You may also wish to test oat-nut bread, a multigrain bread, or any other interesting bread you find at the store or which is home-baked.

Set the plates in an out-of-the-way place. Every day, sprinkle three drops of water on each slice of bread to keep it moist. Write down your observations about each slice of bread every day. How long is it before you see mold forming? Which bread is the first to begin growing mold? Which is the last?

For your report, ask your local bakery if it uses any preservative or mold-inhibitor food additives in

the bread you used in your project. Compare the list
of ingredients of each loaf of packaged bread you
tested.

Results & Conclusion Write down the results of your
experiment. Come to a conclusion as to whether or not
your hypothesis was correct.

Something more

1. Hypothesize that keeping bread cooler will also help
   preserve it. Repeat the above experiment, but place
   an additional slice of each bread in the refrigerator.
   Check all the breads once a day and write down
   your observations.

2. One possible variable in the "something more" experiment above that wasn't controlled was light. The slice of bread *in the refrigerator* did not have light. Was it the temperature or the absence of light that affected the mold test results?

3. If spores get onto the bread from the air, would placing a slice of bread that gets moldy quickly inside a piece of clear plastic wrap keep it from getting moldy?

TICK TOCK TACK
Testing the oxidation time of metals

Purpose To determine which materials become oxidized.

Overview Oxidation is a process that occurs when oxygen combines with various substances. As explained in the last project, oxidation can occur rapidly, like when a candle burns; or take a long time, like when iron rusts.

YOU NEED
- 3 or 4 different kinds of thumbtack
- clear jar
- water
- several days' time
- push pin

If you wished to tack up posters or hang holiday decorations outdoors, you'd want to be sure the thumbtacks or push pins would hold up well in the weather.

Hypothesis Hypothesize which items in water will rust.

Procedure Find several kinds of thumbtacks. Some may be coated, some may be brass. Place the tacks, along with a push pin, in a clear jar filled with water. A mayonnaise, peanut butter or pickle jar will work well. The jar may be plastic or glass.

Draw a diagram on a piece of paper showing the location of each thumbtack and the name of its brand. Set the jar in an out-of-the-way place. Once a day, look at the objects in the jar and write down your observations. The water and the period of time under test is the CONSTANT. The VARIABLE is the different materials being tested.

After one week, examine your notes on the observations you made. Which one started to rust first? Are there any that did not rust? Why do you think some did not rust as much as others?

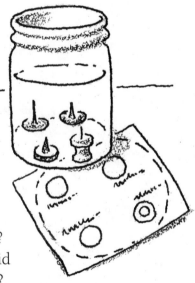

Results & Conclusion   Write down the results of your experiment. Come to a conclusion as to whether or not your hypothesis was correct.

Something more   Try using other types of metal objects, such as staples and brass fasteners. Make daily observations and note any changes.

A PATCH OF WHITE
Green plants and photosynthesis

Purpose Will chlorophyll return to grass that has been deprived of light? Let's figure it out.

Overview Chlorophyll is a chemical in plants that makes them green. With the help of light energy, a plant's chlorophyll turns carbon dioxide and water into sugars it uses for food (the process of photosynthesis) and causes oxygen to be released. If a plant doesn't get the light it needs, photosynthesis can't take place and the plant loses its green coloring. If such a patch of grass is later opened to sunlight again, will the blades of white grass recover…turn green again?

Hypothesis Decide whether you think white blades of grass do recover and turn green again, or if they wither and die and new green grass grows from the roots.

YOU NEED

- a patch a green grass
- 3 foot square piece of plywood
- several weeks' time
- pencil
- paper

Procedure To test your hypothesis, lay a piece of plywood over a section of green grass in a sunny area. Every day, lift up the plywood for a moment and observe the color of the grass underneath. Write down your observations. Is the green color becoming paler? Compare it to the color of the surrounding grass. The patch of grass remains CONSTANT. The VARIABLE is the light reaching the grass.

When the patch of grass has become white-looking, remove the plywood and leave it off. Every day, look at the grass patch and write down your observations. Does photosynthesis begin again in the blades of grass once the light has returned?

Results & Conclusion Write down the results of your experiment. Come to a conclusion as to whether or not your hypothesis was correct.

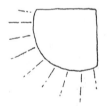

Something more If blades of grass do not recover once they turn white, would they start photosynthesis again if the length of time sunlight was kept from them was reduced? How can you test and show this?

POP GOES THE SODA
Reducing carbonation in soft drinks

Purpose To learn how we can minimize the amount of carbon dioxide ingested from drinking a carbonated soda.

Overview In 1772 in England, Joseph Priestly was trying to discover how to imitate the natural bubbling waters of some mineral springs. That was the beginning of what is done today in many popular drinks. The ingredient that gives soda, sparkling water and other such drinks their bubbly taste is carbon dioxide (chemical symbol CO_2). It makes a soda tickle your mouth and nose when you drink it.

When you breathe out, carbon dioxide is exhaled. Carbon dioxide gas is also formed by burning things that have the element carbon in them, which includes wood, coal and oil. Trees and plants take in carbon

YOU NEED
- 2 20-ounce plastic bottles of soda
- 2 gallon-size sealable clear plastic food bags
- a sunny window
- use of freezer
- clock or watch

dioxide and release oxygen. This cycle is nature's way of providing clean air for people and animals to breathe, as well as carbon dioxide for trees and plants.

Carbon dioxide can be put under pressure and added to drinks to give them a pleasant biting taste and bubbly appearance. When carbon dioxide is added to water it is called soda water. Soft drinks (often called soda pop, or simply soda or pop) have become a big part of our society. They are called soft drinks to distinguish them from beverages containing alcohol (hard drinks). Can you imagine having a barbecue without soda to drink?

Although the carbon dioxide in our drinks is not harmful, the bubbles may cause gas in your stomach,

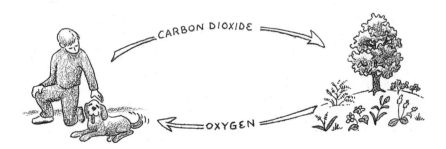

which could cause some discomfort. If you wanted to minimize the amount of carbon dioxide that you ingest, would it be better to drink your soda cold or warm it up, open the cap, then cool it before drinking?

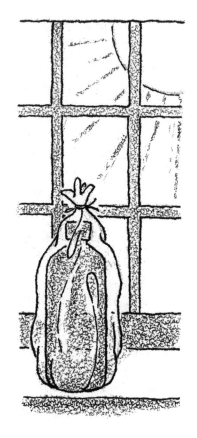

An interesting historical note: Soft drinks are also called "pop" because the bottle caps used prior to the 1890s made a popping noise when removed from the bottle.

Hypothesis Hypothesize which of two bottles of soda will give off more carbon dioxide when opened, as detected by the sound of escaping gas when their caps are finally unsealed.

Procedure Place a 20-ounce bottle of soda in a clear plastic food bag, one that can be sealed. Similarly, place another 20-ounce bottle of the same kind of soda in a plastic bag. Set one soda in the freezer section of a refrigerator. Set the other soda in a sunny window. Wait one hour (no longer!—we do not want the soda in the freezer to freeze).

Place the two sodas on a table, keeping them in the sealed bags. Unscrew the cap off of each by grasping the cap through the sealed bag. The CONSTANTS are the type and amount of soda. The temperature is the VARIABLE.

Which soda made the loudest "hissing" sound as gas escaped when the cap was opened? Which one released the most gas and soda? The soda which made the loudest hiss and pushed out the most liquid now has less carbonated gas in it. If you drank that one, there would be less gas in your stomach. If it was the warm soda, you could now pour it in a glass, add ice to make it more desirable to drink, and it would put less gas in your stomach.

Results & Conclusion Write down the results of your experiment. Come to a conclusion as to whether or not your hypothesis was correct.

Something more

1. Would the color of soda affect results? Repeat the experiment using a clear-colored soda (or use a dark-colored soda if you did the experiment first with a clear-colored soda).

2. Does shaking a bottle before opening it cause it to release more gas when the cap is unsealed? (Do this outside or in a container, so as not to splatter soda on your clothing or other objects around you.)

HOW ELECTRIFYING!
Examining the conduction of substances

Purpose To learn what common materials are good conductors of electricity.

Overview Substances are described by color, texture, smell, feel and hardness. Another characteristic might be an ability to conduct electricity.

Gather some common items (see "needs" list): metal coins made of basic elements like copper, gold and silver jewelry, a plastic comb (or other item made from a chemical process using oil).

A battery is used to store electricity (a chemical reaction takes place within

YOU NEED
- brass fastener
- coins (penny, nickel, dime, quarter)
- plant stem
- twig from a tree
- tea bag
- water
- gold ring or necklace
- piece of silver jewelry
- nail
- plastic comb
- insulated jumper leads with alligator clips at each end
- dual "D"-cell battery holder
- 2 batteries (1.5 volt "D" size)
- flashlight bulb
- bulb socket

the battery). Electricity flows from the negative terminal on one battery to the positive one of another following a path, such as a piece of wire. The material used for the path is called a conductor. An electrical conductor is any material that allows an electric current to flow through it easily. Metal is a good conductor. Copper is used as a conductor in some phone wires, computers and other electronic appliances.

When a conductor provides a path between the negative terminal on a battery and the positive terminal, electricity flows through the conductor. Think of the flow of electricity as if it were flowing water. A metal wire carries the electric flow, just as pipes carry water flow. If we put something in the path of the electric current, the current will try to flow through the object. If that substance is a conductor, the current will flow through. If it is not a good conductor, little or no current will flow through. By placing a light bulb in the path, we can see when current is flowing in the wire, since the current will light the bulb as it flows through it.

Hypothesis Hypothesize which gathered materials are good electrical conductors.

Procedure The battery and light bulb conductivity tester we will construct is the CONSTANT. The VARIABLES are the types of material being tested.

At your local electronic parts store, you should have been able to find two "D"-cell flashlight batteries, a dual "D"-cell battery holder, a flashlight bulb, a socket for the bulb and a set of insulated jumper leads with alligator clips on each end. The alligator-clip leads make it quick and easy to connect all the components together, and to clip onto the different materials we want to test as conductors. Using those alligator clip leads, connect the components together as shown below. Be sure to place the batteries in the holder correctly, each one facing

a different direction. The two batteries used together make 3 volts (each is 1.5 volts and 1.5 + 1.5 = 3).

The battery holder must be one with a wire running from the positive terminal (raised "+" end) on one battery to the negative terminal (flat "−" end) on the other battery to form an electrical path. This connection is made at the opposite end of the battery holder from where the alligator clips are attached. If the holder you have doesn't have this connection, a 3-inch-long strip of aluminum foil ½-inch wide can be placed inside the battery holder, touching the positive end of one battery terminal to the negative end of the other.

The path is completed by clipping a coin between the two alligator clips. Does the bulb light up? Now unclip the coin and, in turn, clip on the other things you've collected (a twig, nail, plastic comb). Make two piles, one of electrical conductors and one of nonconductors.

Note: In a science fair, avoid possible theft from an unattended display by leaving expensive jewelry or coins at home.

Results & Conclusion Write down the results of your experiment. Come to a conclusion as to whether or not your hypothesis was correct.

Something more It's interesting to note that many household items that do not conduct electricity (called insulators) can build up static electricity—a balloon, wool sweater, plastic comb, and glass stirring-rod, for example. Static electricity is a buildup of electrons. Try to build up a static charge on these items. Hold a balloon against the screen of a television set. Take a wool sweater out of a clothes dryer. Run a plastic comb through your hair when it's very dry. Check the items on the conductor tester you constructed in the above experiment and prove that these items do not conduct electric current, even though they can hold a static electrical charge.

WE DEPLORE POLLUTION
Water dilutes, but pollution remains

Purpose   To see if a trace (an extremely small amount) of a solute in a solvent is detectable by taste.

Overview   Just a few drops of an unwanted substance can ruin another substance. A little pollution in pure water can make it undrinkable. Just imagine how few molecules of a substance are necessary for air to carry and spread them. When cookies or a cake is baking in the oven, you can smell it all through the house. Yet, such a small amount of cake is lost to the air that it isn't measurable by us.

For a while, some people thought that unclean water could be cleaned by diluting it (making it weaker). The popular saying was, "The solution to pollution is dilution." Environmentalists now know that just isn't true. That is why everyone must be

YOU NEED
- sugar
- gallon container
- drinking glass
- water
- teaspoon
- hand towel

responsible, so as not to pollute our environment. It may be difficult or even impossible to make water clean and drinkable once certain substances have been mixed into it.

Even if something is diluted, it can still remain contaminated. Even if we cannot taste or see something in the water, that doesn't mean that it is not there. Prove this by using sugar and water.

Hypothesis Hypothesize you can't detect a trace amount of sugar when it is heavily diluted in water. Being aware of this shows how important it is for us to be environmentally wise and keep water from becoming polluted.

Procedure Fill a clean,
empty one-gallon jug
with tap water. The jug
may originally have held milk or
mineral water. Then fill a drinking glass with the
same water.

Add a very, very small amount of sugar to the water
in the jug, only about ⅛ of a teaspoon. Put the cap on
the jug and shake it. Wipe the mouth of the jug with a

towel to be sure there are no granules of sugar on it. Pour a little into a glass and take a sip. Do you taste any sweetness from the sugar you put in? Take a drink of the glass of plain water and compare the taste. The water is CONSTANT. The VARIABLE is the water which contains a trace amount of sugar.

Even though you may not taste any sweetness, the sugar is still present in the water. The sugar is very diluted (weak), but it is there, making the water not completely pure. You know the sugar is in there, because you put it in.

This shows that even though we may not be able to detect the presence of a substance because it is not there in sufficiently large amount, some of the substance could still be present.

Water may meet all of the government regulations for drinkable water, but that water may still have some contaminants in it that may not be healthy for us to drink.

Results & Conclusion Write down the results of your experiment. Come to a conclusion as to whether or not your hypothesis was correct.

Something more Add another ⅛ of a teaspoon of sugar to the jug, shake it and taste it. Continue adding sugar until you can finally taste its presence. How much do you have to add before you can taste the sweetness of sugar?

NO SYRUP

The concept of viscosity

Purpose You may like thick pancakes, while your friend prefers thin ones. Is it possible to control the thickness of pancakes? Let's study the viscosity of pancake batter.

Overview One of the properties of fluids is viscosity. Viscosity is the ability of a fluid to resist flowing quickly. A fluid that flows slowly is said to have a high viscosity; it is "thick." Honey, for example, has a higher viscosity than milk.

Temperature can also be a factor in the viscosity of a fluid (hence the old phrase "as slow as molasses in January"). Gravy that is placed

YOU NEED

- an adult
- instant pancake mix
- butter
- large spoon
- frying pan or skillet
- spatula or wide turner
- 3 dinner plates
- water
- use of stove
- 3 bowls
- 6 pieces of paper
- pencil or pen

in a refrigerator overnight becomes so viscous that it turns gel-like.

Hypothesis Hypothesize that the more viscous the batter, the thicker the pancake will be when cooked.

Procedure Use some instant pancake mix to make pancake batter. Follow the directions for adding water (or milk) and mix or stir as instructed.

Pour the batter into three bowls. Add some water to one bowl to make the batter more watery (lowering its viscosity) and stir.

To the second bowl of batter, add more of the mix powder and stir.

Do not add anything to the batter in the third bowl. Place each bowl on top of a separate piece of paper and write on each paper the contents of the bowl: "standard mix," "more water," "more powder."

With an adult standing by to help, place some butter in a skillet or frying pan, turn the stove burner on to medium heat and spread the butter to grease the pan.

Pour one large spoonful of batter from the standard-mix bowl onto a skillet and cook until golden brown. When done, place the pancake on a plate and lay a piece of paper next to it indicating "standard mix."

In the same way, pour one large spoonful of batter from the "more water" mix on the pan, and cook until golden brown. Lay the finished pancake on a plate and label it "more water." Each pancake should contain the same amount of batter. The heat applied to the pancakes will remain CONSTANT. The viscosity of the batter is the VARIABLE.

Repeat the procedure for the "more powder" batter.

When done cooking, examine each of the three pancakes. Write down your observations. Did one take longer to cook than the others?

Next, taste each one. Is there any difference in taste? Even if they all taste the same, which one is more appetizing to you, a thinner pancake or a thicker one?

If you make pancakes in the future, which will you prefer: making the batter exactly as the instructions recommend on the package, making the batter more viscous, or making it less viscous?

Results & Conclusion Write down the results of your experiment. Come to a conclusion as to whether or not your hypothesis was correct.

Something more

1. How does the amount of time the pancake is cooked affect its thickness, if at all?

2. Does a thicker pancake absorb more syrup than a thinner pancake, making it sweeter when eaten?

FADE NOT
Natural dyes and sunlight

Purpose What is the effect of sunlight on various naturally staining chemical substances, such as those that come from fruits?

Overview Your parents may have been upset when you or someone in your home spilled a certain drink or fruit juice on a good carpet. Some drinks and juices stain. Have you noticed that certain foods even stain your lips and tongue when you eat them? Could fruit or vegetable juices be used as natural dyes for clothing? If so, how do they hold up in sunlight, which often fades the synthetic dyes in carpets and upholstered chairs that are near sunny windows?

Hypothesis Which of the staining liquids in the experiment do you think will resist fading the most? Hypothesize that grape juice (or whichever juice you chose) will resist fading better than the others.

Procedure Take an old white T-shirt, pillowcase, or bedsheet. Check a "rag bag" if you have one, where worn or torn clothing and bedding are kept for use in cleaning or painting jobs. Cut small squares (about 3 inches/7.5 centimeters) of white material.

Place a cereal bowl in a sink. Working in the sink will keep any juice from spilling on the floor. Pour a

little blackberry juice into the bowl. Dip two white cloth squares in the blackberry juice. Lay the squares on an old piece of newspaper to dry.

Rinse the bowl with clear water. Pour in a little beet juice. Dip two white cloth squares into the beet juice and then lay them on the newspaper.

Again, rinse the bowl. Pour in a little purple grape juice. Dip two more pieces of cloth in the grape juice and lay the pieces on the newspaper to dry.

When dry, place one of each colored square in a sunny window (one blackberry, one beet and one grape juice colored squares). Place the remaining three squares in an area away from any direct sunlight. A dresser drawer or on top of the refrigerator would be good places. Can you guess which ones will fade more? Write down your guess to later check and see if you were correct.

After a week or two, compare the colors of the squares that have been in direct sunlight and also the squares that were not exposed to sunlight. Are there noticeable color changes?

The different staining juices are the VARIABLES. The sunlight in the experiment is the CONSTANT, but also in this case a VARIABLE since two pieces of cloth are stained the same—one kept in the dark (the control) and one exposed to sunlight.

Interestingly, some fluorescent-colored bathing suits come with a label attached, warning "Do not expose to direct sunlight for long periods of time!" Does the manufacturer expect people to swim only at night or indoors? Of course, too much direct sunlight isn't good for your skin either!

Results & Conclusion Write down the results of your experiment. Come to a conclusion as to whether or not your hypothesis was correct.

Something more Do your naturally dyed pieces of material lose their color when they are washed?

Look around your house and see if you can find discoloration caused by sunlight in the dyes in carpet and upholstered chairs and sofas. Move small furnishings a little or ask an adult to move something if the object is heavier, and look at parts of a carpet that have been covered (protected from sunlight) for many years. Is the color of the carpet the same on parts that have been largely "in the dark" as on the parts that were in direct sunlight?

THE YEAST BEAST
Fermentation proves yeast is alive!

Purpose When yeast is used in baking, many recipes also call for sugar as an ingredient. The yeast uses the sugar to make carbon dioxide and to give the item to be baked a light and "airy" consistency. But can a sugar substitute be used instead, to activate yeast in baking and have the yeast produce carbon dioxide?

Overview For many hundreds of years, people baked bread,

YOU NEED

- active dry yeast (available at the grocery store)
- 3 clear drinking glasses
- warm tap water
- measuring cup
- natural white sugar
- brown sugar
- sugar substitute (saccharin)

using yeast as an ingredient, without knowing just why it makes bread dough bubble and rise. They thought it was simply some sort of chemical reaction. It took Louis Pasteur and other scientists doing experiments in the 1850s to prove that the yeast ingredient was actually a living organism. It is this organism that causes the chemical change in bread dough.

Yeast digests sugar and starch and turns them into alcohol and carbon dioxide gas. This breaking down of sugar and starch is called fermentation. The carbon dioxide gas bubbles up through the bread dough, making the bread rise higher and become more porous (full of tiny holes). Although fermentation is also used in making alcohol, we don't taste alcohol when we eat bread with yeast, because any alcohol that is produced evaporates while the bread is being baked.

Does fermentation also take place when a sugar-substitute product is used in baking with yeast? This is an especially

important question if someone wants or needs to bake sugar-free foods and is considering using a sugar substitute in a recipe.

Hypothesis More gas is released when yeast is mixed with water and sugar than if it is mixed with water and a sugar substitute.

Procedure Pour a little active dry yeast into each of three clear drinking glasses. Pour just enough to cover the bottom of each glass completely.

Add one level teaspoon of natural white sugar to one glass, one level teaspoon of brown sugar to another glass and one level teaspoon of an artificial sweetener (such as saccharin).

Measure and pour ¼ cup of warm tap water into each glass and swish the glass around, making a circular motion with your hand, to gently stir the contents. The amount of yeast and water is kept CONSTANT. The VARIABLE is the type of sweetener that is used.

Set each glass on a sheet of paper and write on the paper the type of sugar that the glass contains. When doing science experiments, it's important to label or identify each container as you do your experiments, in order to avoid confusion later.

Observe the glasses for a little while, watching for a foam of bubbles to appear, indicating the presence of carbon dioxide gas. Which one foams up the most and thus has the most response to the yeast?

Results & Conclusion Write down the results of your experiment. Come to a conclusion as to whether or not your hypothesis was correct.

Something more Experiment checking the reaction of yeast to other types of substances that contain sugar (honey or pancake syrup, for example).

BETTER BUBBLES

Safe and natural monster-bubble solutions

Purpose Everybody likes to make bubbles—the bigger, the better. But is your bubble solution toxic or entirely safe?

Overview Bubbles form in water when air is trapped. But bubbles made of water alone cannot be very large or survive in air. Something must be added to water to chemically change it so that water molecules will hold more tightly together (called surface tension).

YOU NEED

- confectioners' sugar with cornstarch
- warm water
- measuring cup
- liquid dish soap
- tablespoon
- a bottle of "bubble stuff" from the toy store
- a bubble wand (from store or homemade)
- bowl

When you buy "bubble stuff" in a toy store, an ingredient may have been added to make larger and longer-lasting bubbles (possibly glycerin) but the solution can be toxic, or poisonous if swallowed.

Hypothesis Hypothesize that it is possible to make a monster-bubble solution that is safer, so that even little kids can enjoy making big bubbles.

Procedure Pour ¾ cup of warm water into a bowl. Add 4 tablespoons of liquid dishwashing liquid. This liquid soap is an "emulsifier" that will help the molecules of water hold together. Stir in 2 level tablespoons of confectioners' sugar (which contains cornstarch).

Now make bubbles with the store-bought "bubble stuff," using the wand that comes with the bottle. Then, rinse off the wand and use it to make bubbles with your homemade bubble solution. The wand used to create the bubbles is the CONSTANT. The bubble solution is the VARIABLE. (You can use a pipe cleaner or some other wire bent in the shape of a wand, too, but you must use it the same way for both bubble solutions because it must remain constant for the project.)

Does your homemade bubble solution make as many bubbles and as large bubbles as the store-bought bubble liquid? If so, you have made a safe bubble toy. Of course, you still don't want to swallow your homemade soapy liquid, but it would not harm you if you did.

Results & Conclusion Write down the results of your experiment. Come to a conclusion as to whether or not your hypothesis was correct.

Something more

1. How big can you make a bubble using your bubble solution, before the water molecules can't hold together? How long can you get a bubble to last before it pops? Why does it pop (could it be due to evaporation)? Will a bubble last longer in higher humidity (in a steamy shower, for example)?

2. Is there any difference in using warm water or cold water in making your homemade bubble solution?

3. Will the kind of water matter—for example, tap water, distilled water, spring water? Does the "softness" or "hardness" of water make a difference? "Hard" water has more minerals, which usually makes it more difficult to make suds.

4. Experiment by substituting other safe ingredients. Instead of using confectioners' sugar, try honey or maple syrup as a thickening agent, but only use a very small amount.

COLORFUL DISGUISE

Smell and taste: team players

Purpose How important is the sight of a food in its identification, compared to smell and taste?

Overview Our senses of smell and taste are chemical processes that our bodies use as a team to help us evaluate foods. But do you think it is easy to identify a food from taste only, or in combination with smell, without the aid of the sense of sight?

   To test this, we want to prepare pieces of several fruits and vegetables for various friends and others to sample a taste. And, since they may be able to identify even a small piece of fruit or vegetable by sight (and we'd

YOU NEED

- an adult
- several friends
- apple
- pear
- cucumber or squash
- cantaloupe or other melon
- knife
- 4 bowls or containers
- aluminum foil
- blue food coloring
- eye-dropper
- paper and pencil

rather not subject them to being blindfolded), we will disguise each food sample so it cannot easily be identified.

To carry out the experiment, we'll need to use fruits and vegetables with similar textures. Then we'll cut the pieces into very small cubes and change the normal "look" of the food with food coloring, so its color will not be a help in guessing a food's identity.

Hypothesis Your friends will make more incorrect (or correct) guesses than correct (or incorrect) guesses based on taste alone...or based on both taste and smell.

Procedure Have an adult help you cut the *fleshy* part of an apple, a pear, a melon such as a cantaloupe or honeydew, and a cucumber into a number of small cubes. Be sure not to include any seeds or skins, which might be a help in identifying the food.

Place apple pieces in one bowl, melon in another, pear in a third and cucumber in a fourth bowl. Write the name of each test fruit or vegetable onto a piece of masking tape and stick it onto the bottom of the proper bowl of test samples.

Now take a single piece of each fruit or vegetable and place it in a small section of aluminum foil and close it up loosely. Place it alongside the bowl of fruit samples. Disguise the remaining cubes of food in the bowl with blue food coloring (except for berries, blue is not a common fruit or vegetable coloring).

Place the sampling bowls at least a few feet apart and, one by one, have a few of your friends approach a bowl. While holding their nose tightly, have them take a cube of food, taste it and guess what they think it is…*by taste alone.* Then let them sniff the bowl of colored fruit pieces, thereby adding the sense of smell, and allow them to change their taste-only guess if they wish to. Once each friend finishes the test, let him or her see the normal-colored fruit pieces that you placed in each foil packet. Are your friends now able to recognize the fruit? Keep track of all their guesses. You can use the information to make up a chart for your project display.

The senses used by each individual taking the test and the fruits used remain CONSTANT. The color of the fruits are the VARIABLE from their natural color. Tally how many guesses were right after tasting and after smelling (and seeing with normal coloring) and how many were wrong.

Results & Conclusion Write down the results of your experiment. Come to a conclusion as to whether or not your hypothesis was correct.

Something more

1. Do the above experiment, first testing five kids your age and then five adults. Do you think adults will have more correct guesses than your friends?

2. Select other fruits and vegetables that have similarly fleshy parts so they could be used in this type of taste test. How about squash and tomato? What about the strength (sweet, sour, bitterness) of the flavors?

3. In addition to fruits and vegetables, do a similar experiment using cubes of lunch meat, for example, bologna and turkey breast, or different cheeses.

SCENT IN A CUBE
Releasing fragrance with heat

Purpose You know that flowers outside give off fragrances and will do so in your home, if you bring them inside. But how can you make something in your home give off more of its scent?

YOU NEED

- an adult
- bar of soap with a fragrance
- knife
- microwavable dish
- use of a microwave oven

Overview Smell is a chemical process that takes place as tiny invisible molecules leave an object and travel through the air

to reach our nose, where a chemical solution and nerve cells line the inside of the nose to detect the smell.

Smells can remind us of special times and events. The smell of cedar may make you think of a cedar chest someone in your family uses to store clothing. If you decorate a live tree at Christmas time, you may get the feeling of that season when you smell that particular kind of tree. Think about what odors remind you of a happy time at home during a holiday, or a visit to grandma's house. Many people like the smell of

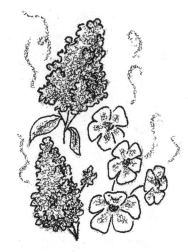

peppermint, apple, pine and cinnamon. They buy candles that are scented, which give off their aroma when lit. They buy incense burners to spread an aroma throughout their room. There are small potpourri pots where water and pieces of fragrance are heated with a small candle underneath.

Heat can cause a material to release molecules into the air and be diffused (going from an area of more concentration to an area of lesser concentration). Then we can detect the airborne molecules with our nose. That is why a scented candle can be in a room, but you

won't smell it until it is lit, and the potpourri pot does not release its scent until it is warmed.

Hypothesis Hypothesize that heat will cause the fragrance in a perfumed soap to be released and fill the air in a room with an aroma.

Procedure Have an adult carefully cut a small cube from a bar of soap that is perfumed or has a fragrance. Notice that you probably do not notice the scent very much unless you place the soap fairly close to your nose. The soap is CONSTANT. The VARIABLE is the heat that is applied to the soap.

Place the soap on a microwavable dish and place it in a microwave oven. Heat it for several seconds. Microwave ovens have different powers, so you may have to experiment with the amount of time. Start with about five seconds.

Open the door of the microwave, but *don't touch the soap*; it may not *look* hot to you, but it might still be hot enough to burn. After a few minutes, walk around the room and sniff the air. Can you detect the soap's smell in all areas of the room? How about in the next room?

Results & Conclusion Write down the results of your experiment. Come to a conclusion as to whether or not your hypothesis was correct.

Something more

1. Can the same piece of soap be reheated again and again to release even more fragrance?

2. Purchase two identical car fresheners, the small decorative fresheners made to hang from the rear-view mirror. Put one in a sealable plastic bag and place it in a refrigerator freezer. Place the other in a sunny window. Later, remove the one from the refrigerator and open the bag. Is its smell as strong as the freshener in the sunny window? If the freshener in the sunny window has a stronger smell, does this mean that car fresheners give off more scent in the summertime than they do in winter?

HAVE A TASTE, BUD
Sugar sweetness a matter of chemistry

Purpose We experience the sense of taste when we eat or otherwise put things in our mouths. What do we taste and why do we taste it? Let's find out.

Overview The sense of taste is a chemical action that takes place on our tongues. Test that taste is a chemical process by laying a dry cornflake on your tongue. It won't have any taste at all until your saliva begins dissolving it. Do the same with a dry cracker.

YOU NEED

- cornflakes
- crackers
- peppermint stick hard candy
- piece of licorice
- cherry lollipop
- grape lollipop
- lemon lollipop
- glass of water

Because food must be dissolved before it can be tasted is proof that taste is a chemical process. It is only possible for us to taste foods that are in a liquid state. If they're not, the saliva in our mouth dissolves the food, turning it into a liquid.

Small organs, called "taste buds," on our tongues then evaluate the dissolved food and give us the sensation of taste. Four sets of taste buds are used to detect the sweet, sour, salty and bitter taste groups. As shown, these four kinds of taste buds are located at different areas on top of the tongue.

You can easily test this; open your mouth wide and briefly rub a piece of peppermint stick against the back part of your tongue, being careful not to touch the front part. Don't place the candy too far back on your tongue or it might make you gag. After a brief rubbing, notice any taste in your mouth. Now rub the peppermint on the front area of your tongue. Can you taste the sweetness of the candy better?

Hypothesis Hypothesize that the organs that cause you to be able to taste and isolate flavors are grouped on top of the tongue.

Procedure Place a hard candy peppermint stick under your tongue and a small piece of licorice on top of your tongue. Wait a few moments for your saliva to dissolve some of the candies. Which candy do you taste?

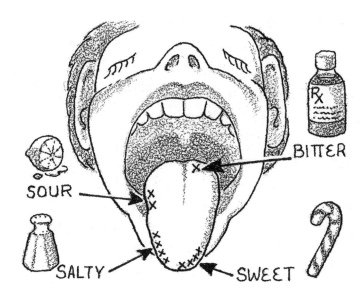

SOUR

BITTER

SALTY

SWEET

Take a sip of water to clear the taste from your mouth (official tasters call it "clearing the palate"). Now reverse the candies, placing a licorice piece under your tongue and touching the peppermint stick to its top. Which candy do you taste now?

Experiment further by placing various flavors of lollipop under your tongue and onto areas of its surface; try cherry, grape, lemon, lime and other flavors that you know are easy to identify. Your tongue with its taste buds is the CONSTANT. It's the different *areas* of the tongue undergoing testing that are the VARIABLES, as are the different lollipop flavors.

Results & Conclusion Write down the results of your experiment. Come to a conclusion as to whether or not your hypothesis was correct.

Something more Prepare small pieces of fruit: orange, apple, grape, peach and others. Have friends close their eyes while you lay a piece of fruit on their tongues. Can they identify the fruit? Can they tell the difference between sweet, sour, salty and bitter?

BUILDING BLOCKS
Growing natural crystal structures

Purpose Certain substances just naturally take flat-sided crystalline shapes. Here's how to watch, as you "grow" your own crystals.

Overview Everything is made up of small particles called atoms and molecules. In some solid substances, these atoms and molecules are arranged together to make three-dimensional patterns, which are repeated over and over until they are big enough for us to see with a microscope or a magnifying glass. These substances are called "crystals." Crystals have shapes that are characterized

YOU NEED

- an adult
- 1 box of lemon gelatin dessert powder
- red food coloring
- kitchen measuring cup
- 10 plastic disposable spoons
- 10 paper cups
- water
- pencils (or pens)
- 10 small notepad-size sheets of paper
- a cooking pot
- use of stove-top burner
- use of refrigerator

by their smooth, flat surfaces with sharp edges. Crystal-shaped substances include sugar, table salt, gold, silver, topaz, quartz and copper sulfate.

Although non-living things do not grow, the molecules of crystals can pile together to "grow" bigger in size. When enough sugar is dissolved into water, sugar crystals will build up or "grow" on a piece of thread or string that is left in the solution until a clump big enough to see with the unaided eye appears.

Examine a few grains of sugar under a microscope. If you do not have a microscope, you can use a magnifying glass that has a high magnification. Observe that the tiny grains of sugar have a cube- or block-like shape. This is one kind of crystal shape.

Examine a few grains of table salt under a microscope or a

magnifying glass of high magnification. Observe that salt, too, has a cube-like shape and therefore it is also called a crystal.

Hypothesis Hypothesize that, since you have observed salt to be a crystal, it too can be made to accumulate (build up) and "grow" a large crystal object.

Procedure Fill a short drinking glass with hot water from the tap. Be careful not to burn yourself! Add a spoonful of sugar to the water and stir. Continue to add one spoonful at a time and stir until no more sugar can be dissolved in the water. You can tell when this happens, as sugar will begin to build up at the bottom of the glass and will not dissolve, no matter how long you stir. When a solution is holding as much of a substance as it can, it is called a "saturated solution." A "solution" is a solvent (the material that you use to dissolve) that contains a solute (the material that gets dissolved).

Tie a piece of thread or thin string onto a pencil at its center. Lay the pencil on top of the glass of sugar water and let the string hang down into the water. Set the glass in an out-of-the-way place, where it will not be bumped or moved. Wait two to three days.

Then observe the buildup of sugar crystals on the string. We are "growing" a big crystal from many small crystals.

Fill another short drinking glass with hot water from the tap. Use caution handling hot water so as not to burn yourself. Stir in a spoonful of table salt. Continue to add salt and stir until the solution has become saturated (no more salt can be dissolved and excess salt can be seen at the bottom of the glass). Tie a piece of thread or thin string onto the center of a pencil. Lay the pencil on top of the glass and let the string hang down into the solution.

The water and string are CONSTANT. The solute (the material being dissolved) is the VARIABLE (sugar and salt). After two or three days, examine the string. Have crystals of salt begun to build up on the string? Was your hypothesis correct?

Results & Conclusion   Write down the results of your experiments and come to a conclusion as to whether or not your hypothesis was correct.

Something more

1. Can you build your crystal objects even bigger by adding more sugar and salt to the solutions and waiting a few more days? Do not heat the water again, because you don't want to dissolve the crystals that have already formed on the string. You want to add to them.

2. Ice can form crystal structures, too. On a morning when frost makes designs on your house windows or car windshield, use a magnifying glass of high magnification to examine the frost for evidence of crystal shapes.

3. Honey can crystallize and turn into a solid; but can it be restored to a liquid form by warming it?

TRICKING THE BRAIN
When a food's color is changed

Purpose To determine how important color is in our expectations of the taste of a food.

Overview From our life experiences, we learn to expect something that tastes like lemon to be yellow in appearance. We expect something that tastes like cherry to be red in color. Does color really play a big part in how we expect a food to taste?

YOU NEED
• an adult
• salt
• sugar
• 2 short drinking glasses
• thin string or thread
• spoon
• hot water from the tap
• 2 pencils
• magnifying glass (microscope preferable)

In this project, a batch of lemon gelatin will be made, with a spoonful given to each of 10 people. Each person will be asked to take a spoonful and try to identify the flavor. The lemon dessert, however, will not be yellow!

Hypothesis Most people will not be able to recognize the true flavor of a gelatin dessert if the natural color of it has been changed.

Procedure Following the instructions on a package of lemon-flavored gelatin, ask an adult to help you make a bowl of the dessert. (For safety, it's important that an adult stand by whenever you work around or use a stove.)

Before you place the lemon gelatin liquid in the refrigerator (be careful of splashes), add several drops

of red food coloring and stir. Add the food coloring until the lemony-yellow liquid has turned a deep red.

When the disguised lemon gelatin has cooled and hardened, place a spoonful into each of 10 small paper cups. Give 10 people (friends or family members) a cup with a gelatin portion and a spoon. Ask them to each taste the dessert and, without speaking about their choice, to write on a piece of paper the flavor of the gelatin. (Be sure your test subjects aren't being influenced by seeing others' answers.)

Ten people are being asked to give us an idea of how most people will probably answer. Ten is our "sample size." We are using a few people to estimate how a lot of people may respond to the test. A sample size is when a smaller group is being tested which, if large enough, will hopefully give us a true picture of a larger group.

When you have tested 10 people and collected your data, determine the percentage of people

$$8/10 = .8$$
$$.8 \times 100 = 80\%$$

who were fooled by the unexpected color. To find percentage, simply divide the number of wrong guesses by 10 and multiply the answer by 100. If, for example, eight people guessed wrong, that would mean 80% of those tested were unable to correctly identify the flavor.

Results & Conclusion   Write down the results of your experiment. Come to a conclusion as to whether or not your hypothesis was correct.

Something more

1. Test the concept of "sample size" by testing 20 people (you only need to ask 10 more people and add those results to the first 10). Compare the percentage of wrong guesses to 20. Is the percentage about the same as it was with only 10 people? Do you think that a sample size of 10 people used in the original experiment was enough to get an accurate result?

2. Are some flavors easier for people to identify, even if the color is different? Try a common flavor, for example orange, and add red food coloring until it is red. Then ask people to identify the flavor. Are more people able to correctly identify orange than lemon, even though both are disguised by a strange coloring?

YOU'VE CHANGED!
Identifying chemical and physical changes

Purpose To understand the difference between a physical change and a chemical change.

Overview Some changes which take place in substances are "physical changes" and some are "chemical changes." What is the difference?

When a chemical change takes place, the substance often takes on new properties. Glass is made from sand and limestone, but has properties unlike either sand or limestone. Carbon, hydrogen and oxygen can be combined to form table sugar, a compound that is nothing like the elements that make it up. Hydrogen and oxygen by themselves burn, but when two atoms of hydrogen combine with

YOU NEED
• drinking glasses
• water
• ice cubes
• 2 slices of bread
• use of a toaster
• pepper
• fruit (for fruit salad)
• fresh milk
• sour milk
• unbaked cookie dough
• baked cookies

one atom of oxygen it forms a molecule of water, which not only doesn't burn, but can be used to put out fires!

It is often difficult or impossible to "undo" the results of a chemical reaction. However, a physical change can usually be reversed.

For example, heat may cause a chemical change. Compare a slice of bread to one that has been toasted. The heat has caused a chemical change in the bread, and we cannot restore the bread to the way it was before it was toasted. A log burning in a fireplace is having its chemical composition changed to carbon, and other chemicals in gas form, while giving off heat, light and sound. It would be impossible to combine carbon, and gases, and remove heat to restore the log to its former state.

Water can be changed by temperature to take the form of a gas, a liquid, or a solid (ice). This is not a chemical change; it is a physical change. Ice can easily be turned back into a liquid, and the liquid retains all of the same properties it had when it was previously a liquid. This is a case where heat causes a physical change but not a chemical change.

Hypothesis By gathering together some common everyday materials, it is possible to clarify and display the results of three physical changes and three chemical changes.

Procedure Set up the following examples of change and explain the differences:

Physical change:

1. a glass of water and ice cubes

2. a glass of plain water and a mixture of water and black pepper (pepper does not combine with water, and can be easily removed)

3. cut-up fruits and a fruit salad (even though fruit pieces are mixed together, you can still pull out individual slices to separate them again)

Chemical change:

1. a slice of bread and a slice of toasted bread

2. unbaked cookie dough and a baked cookie

3. fresh milk and sour milk

The foods you start with, before any changes are made to them, are the CONSTANT.

The physical VARIABLE with water and water containing pepper is the addition of pepper; with water and ice cubes it's temperature; with cut-up fruit and fruit salad it's the mixing.

The chemical VARIABLE with bread and toasted bread and with unbaked and baked cookie dough is the addition of heat; with fresh and sour milk it's the souring.

Results & Conclusion Write down the results of your experiment. Come to a conclusion as to whether or not your hypothesis was correct.

Something more

1. Can you name some physical changes that, at first, appear to be chemical changes. For example, adding powdered drink mix to water. It is physical, because the water can be separated from the mix by evaporation (leaving the mix behind). Baking cookies is a chemical change. Breaking the finished cookie in half to share with a friend is a physical change.

2. If newspaper is placed in a sunny window for several days, it changes color. Is that a physical or chemical change? Can it be easily reversed?

3. Is rotting fruit a chemical or a physical change?

ENERGY

ENERGY

Energy

Welcome to the fascinating world of energy! This section explores projects in energy and the physics of energy. The term "energy" is difficult to give a meaning to, since it is found in many forms and is closely linked to "forces" (magnetism, gravity, wind, etc.). Physicists define energy as the ability to do work, and they define "work" as the ability to move an object over a distance.

Forms of energy include solar, mechanical, chemical, electrical, moving fluids (both gases and liquids, etc.), heat, light, sound, pressure, thermal, nuclear, electromagnetic waves, respiration (living things get energy from foods, and muscles do work), and the forces of weather, gravity, and magnetism.

Energy can be transferred from one object to another; a rolling marble strikes a stationary marble and causes it to start rolling. Energy can be converted from one form to another, such as light energy to heat energy. Albert Einstein is known for the formula $E = mc^2$ he put forth in 1905, stating that matter can be changed into energy and energy into matter.

Energy is said to be either "kinetic" or "potential." Potential energy is "stored-up" energy—something that has the ability to do work. Kinetic energy is the energy of movement, when work is actually being done. Potential energy can be converted into kinetic energy, and vice versa. Energy from sunlight is stored in trees (potential energy), which can be burned in a fireplace to produce heat (kinetic energy). Roll a rock up a hill (using kinetic energy), and set it on the hilltop (potential energy), where, because of gravity, it has the potential to do work (when it falls).

This section is a little different—instead of laying the experiment out for you, it gives some ideas and basic projects. You can take these and build on them to create your own individual experiment—remember to follow the scientific format that outlines the experiments in the other two sections.

SHRINKING CUBES
Changing the sun's light into heat energy

Imagine a sunny day at a picnic. You pour a glass of cola soda to drink. Your friend fills a glass with a lemon-lime soda he likes. You both take one ice cube. Which ice cube will last longer?

Sunlight turns into heat energy. Things that are dark in color absorb more light energy than those that are lighter, so they become hotter. Will the darker soda collect more sunlight and melt the cube faster?

YOU NEED
- 2 same-size clear drinking glasses
- clear-colored soda drink
- dark-colored soda drink
- a table by a sunny window
- 2 same-size ice cubes
- a dark room

Take two same-size clear glasses, fill one with a clear or light-colored soda and one with a dark-colored soda. Fill each glass to the same height, not quite to the top. Place the glasses in a sunny window for a half hour. Then, take two ice cubes of equal size and drop one into each glass. Which of the two ice cubes lasts longer? Why?

In this project, an assumption is made. We are assuming (we "think") that the kind of soda itself (flavor, sweetness) does not affect the melting of the

ice cubes. To prove that our assumption is correct, do the experiment again. This time, set both glasses of soda in a dark room instead of in the sun. If both ice cubes take the same amount of time to melt, then the sodas had an equal effect on the ice cubes, and our assumption is correct.

Test out other drinks: orange juice, red punch, lemonade, ginger ale. Try carbonated/noncarbonated, diet (sugar substitute)/high sugar, with/without solids (pulp), etc.

What else about a drink might affect the melting speed of ice cubes? How can you find out if it does?

FROSTY'S SUNSCREEN
Warding off the sun's heating rays

It's fun to build a snowman and have it stand guard in your yard all winter long. But rising temperatures and the sun's heat are not kind to snowmen. It can quickly make them melt away.

YOU NEED
- a sunny day with snow on the ground
- large black plastic bag
- large white plastic bag

Will putting a "hat" or kerchief on your snowman's head help shade him from the sun and keep him around longer?

On a sunny day when there is snow on the ground, build two identical snowmen. Fold a large black plastic (trash) bag into a kerchief or hat and place or tie it on the head of one snowman. You might need to use snow or small twigs to help hold it in place.

Fold a large white plastic bag as you did the black and place it on the head of the other snowman. Again, keep it in place by tying or using snow or small twigs.

As the day goes by, check each snowman to see if there has been any melting. If so, which one's head shows the most melting?

Plastic bags often come in other colors—blue, red, and green, for example. Would using these colors as hats make any difference in keeping a snowman around longer? Would using no kerchief make a difference? (If there's not enough snow available to make several large, whole snowmen, just make large snowball "heads" and wrap same-size sections of the different-colored bags on them for this experiment.)

GETTING STEAMED
Water vapor put to work

Steam is water changed into a gas by heat energy. We use the energy of steam to do many things. Steam has been used to power boats and trains. Steam turbines generate electricity when steam pressure pushes against blades or paddles connected to a shaft and turns the shaft. On the other end of the shaft is an electrical generator.

YOU NEED

- an adult
- toy pinwheel
- Pyrex beaker
- stove burner or hot plate
- one-hole rubber stopper
- glass tube with a 90° bend

Get a Pyrex beaker, a one-hole rubber stopper, and a glass tube with a 90-degree, or right-angle, bend. These items can be purchased inexpensively at a science store or borrowed from your science teacher at school.

Never work around a hot stove without an adult with you. Be very careful! The stove burner, the beaker, and the escaping steam will be hot. Do not touch them!

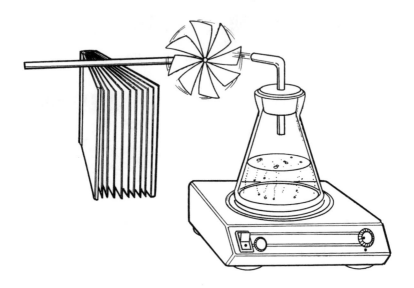

Pour some water into the beaker. Insert the rubber stopper in the top; then ask an adult to gently push one end of the bent glass tube through the hole in the stopper.

Set the beaker on the burner of a stove or a hot plate and turn it on high heat. Open the pages slightly of a tall hard-bound book and stand it next to (but not too close to) the burner. Lay a toy pinwheel on a stick on top of the book and extend it out so that the pinwheel paddles are in the path of the escaping steam from the tube. Do you think heat energy is being changed into mechanical energy?

HOT STUFF

Heat energy from decomposition

Heat energy is given off when organic things (material that was once alive) decay. Many people who plant gardens have "compost piles" to make fertilizer for feeding the garden. A compost pile is a small area, often boxed, filled

YOU NEED

- freshly mowed grass clippings
- a warm, sunny day
- lawn rake
- clock or watch

with dead or dying plant and animal leavings such as peels and scraps from the kitchen, fallen leaves, grass clippings, manure, hay, and other things that rot. The material is stacked and allowed to decay for months, as it turns into rich fertilizer. As it decays, heat is given off.

When your lawn or your neighbor's lawn is mowed, gather some grass clippings by using a lawn rake. Make a pile of grass clippings 1 foot (30 cm) high and about 1 foot (30 cm) in diameter. Place the pile on the lawn in a bright, sunny spot. Let it sit in the sunlight.

After two hours, use the rake to make another pile of grass clippings and place it next to the first one. Both piles should be the same size. Wait ten minutes. Then push one hand into the middle of each pile. Does the inside of one pile feel warmer than the other? If so, which pile feels warmer, the one you just raked or the one that has been sitting in the sunlight for two hours?

Do you think grass clippings can be used to make a good habitat or nest for some animals?

ROLLING STOCK
Potential energy, mass, and gravity

Energy used to move an object up to a height is stored in the object as "potential energy" because gravity pulls downward on the object and will cause it to move. If two objects are raised to the same height, which has more energy stored in it (which requires more energy to move), the lighter object (less mass) or the heavier one (more mass)?

YOU NEED

- 2 two-liter plastic soda bottles
- books
- ruler
- 2 boards, about 1 x 4 feet (30 x 120 cm)
- water
- an adult

Stack several books on the floor, making two piles the same height, about 1 foot (30 cm) tall. Make two ramps by propping one end of each long board up on a stack. Shelf boards work well if you have them; if not, have an adult help you find two same-size boards, or cut two boards from a section of plywood.

Fill a plastic two-liter soda bottle with water and screw the cap on tightly. Screw the cap onto another, empty, plastic two-liter soda bottle. Lift both bottles to the top of the ramps, laying them on their sides, and hold them. Then let go of both of them at the same time. Which one travels farther? The one that travels farther had more potential energy, and therefore also took more energy to move to the top of the ramp.

The bottle that travels farther also has more "momentum." Momentum is a force that moves an object. It is the product of mass times velocity. The bottle filled with water has more mass than the empty one. Now, why do you think it is hard to stop a moving train quickly?

A ramp is an "inclined plane," a type of simple machine. Research inclined planes.

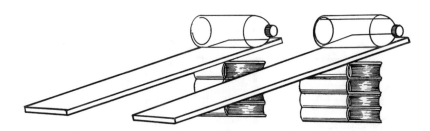

SAILS ALOFT
Using wind energy to power a boat

From mankind's early days, sails have been used on boats to harness the energy of the wind. Let's make small sailboats, using quart-size milk cartons, and experiment with different shapes and sizes of sails.

Lay an empty quart milk carton on its side, with the spout-opening upward. With scissors, cut off the top half of the carton lengthwise to make a boat.

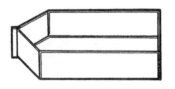

Near the front (pointed end), place a small mound of modeling clay. To make a mast, push the eraser end of a pencil into the clay. A small amount

of clay may be needed
near the back of
the boat to
keep the
boat
balanced. Tape a
1-foot-long (30 cm)
piece of string onto the
back of the boat so it will drag in the water
(this will help hold the boat on course).

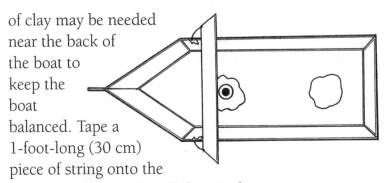

Cut a sail out of a piece of paper and tape it to the
pencil mast. You may need tape or thread to hold the
bottom ends of the paper to the sides of the boat,
keeping the sail tight in strong winds. Make several
boats using sails of
different sizes and
shapes.

Find a place
outdoors—a small
lake, shallow pond,
wading pool, public
fountain—where
you can sail your

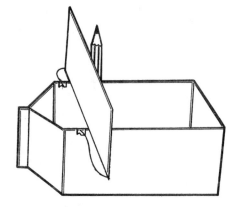

boat. Have an adult with you for safety around water. Which sail design do you think will make the boats go fastest? Test your sail design.

At the library, research "sailboats" and try different designs on your boats.

SALT OR NOT?

Comparing solar energy storage in salt and fresh water

Compared to air, water is slow to change temperature. If the weather has been hot for a few days and the water in a swimming pool is warm, one night of cooler temperatures will not change the temperature in the pool very much. It will still be almost as warm the next day.

YOU NEED

- salt
- 2 two-liter soda bottles
- 2 thermometers
- modeling clay
- hot tap water
- clock
- paper and pencil

Does the ability of water to hold the heat energy it has collected differ if the water is salt water or fresh water? Does a saltwater lake cool differently than a freshwater lake when the sun sets?

Fill 2 two-liter soda bottles with equally hot water from your kitchen sink (let it run a bit until you get a constant temperature). Add eight teaspoons of salt to one bottle. Stir to dissolve the salt thoroughly.

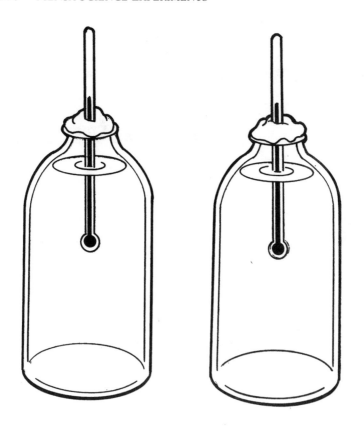

Instead of screwing the caps on, lightly place a ball of modeling clay over the mouth of each bottle. Carefully push a thermometer into the bottle through the clay, so that the bulb of the thermometer is in the water. Make sure you can read the temperature on the thermometer, then press the clay against the thermometer to hold it in place. Do this to both bottles.

Every ten minutes, read the thermometers and write down the temperature readings for both bottles. After two hours, compare the temperatures you recorded. Did they both lose heat energy at the same rate?

THINGS ARE HEATING UP
Graphing solar energy collection in materials

As the sun beats down on the Earth, heat energy is absorbed by everything on the surface. What gathers more of the sun's heat energy: air, water, sand or stone?

Find four large glasses or, using scissors, carefully cut the top half off four clear 2-liter plastic bottles. (Place the discarded tops in a recyclables trash container.) Stand a ruler upright on a table alongside the container (glass or bottle bottom). Four inches (10 cm) above the table surface, mark the container using a small strip of adhesive tape. Place the top of the tape at the 4-inch height. Do this for all four containers, then write the contents of each container on the tape (air, water, sand

YOU NEED
• scissors
• 4 large glasses (or empty 2-liter bottle bottoms)
• a sunny window
• ruler
• masking tape
• water
• sandy soil
• small stones
• 4 thermometers
• 2 pencils or sticks
• clock
• paper and pencil

or soil, stones). The tape will serve as both a label and the fill-to mark for the containers.

Leave one container empty ("filled" with air). Fill another to the 4-inch mark with water. Fill a third up to the mark with sandy soil. The bulb of each thermometer must be placed, hanging, in the middle of each container, not touching the sides or bottom. For the containers of air and water, suspend the thermometers by placing a section of tape at the top of the thermometer and over a pencil or stick placed across the top of a container. The thermometer bulb should be hanging

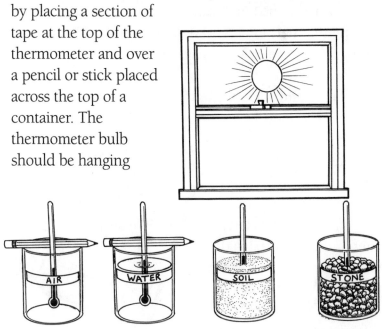

about 2 inches (5 cm) from the bottom of the container.
Carefully push a thermometer down partway into the
sandy soil. For the fourth container, hold a thermometer
inside the container with one hand so that the bulb is 2
inches (5 cm) below the fill line. With the other hand,
gently place small stones, about ½ to 1 inch (1–2.5 cm)
in diameter, up to the container's fill line.

Set the four containers in a sunny window. After one
hour, measure the temperature of each material by
reading the thermometers. Write down the
temperature readings and compare them.

Take all four containers out of the sunny window
and place them somewhere in the room that is out
of direct sunlight. Every five
minutes for thirty minutes,
read the temperatures
on the thermometers
and write them down in a
list. Thirty minutes later,
after an hour has passed,
record the final temperature
readings. Which container
lost heat the fastest? Which
material continued to give off

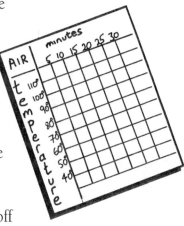

heat the longest? Was this also the same material that gathered the most heat?

For your project display, make up four simple graphs; see sample for *Air*. Across the top (X-axis) put in the time of the readings (every 5 minutes). List temperatures going down the chart (Y-axis). Then enter your own time–temperature readings of each tested material on its chart.

ONE IF BY LAND
Comparing land and water solar-heat storage

The sun warms our planet. As the Earth turns and night falls, the surface the sun has warmed begins to cool, radiating the heat energy absorbed during the day. What loses heat faster— a body of water or the land near it?

Find a lake, pond or large swimming pool. Always have an adult with you for safety when you are working around deep water.

Tie string to each thermometer. At the end of a sunny day (about 5:00 p.m.), poke a hole 1 to 2 feet deep (30–60 cm) into the ground near a body of water with a long stick. A stake from a game of horseshoes or croquet works best. Lower one thermometer into the hole and tie the other end of the string to the stake. Then, hold the other thermometer in the water. Several minutes later, pull up both thermometers and read and record the temperatures.

Take temperature readings every hour as the sun goes down and record your measurements on a chart. Compare the changes in temperature over time. Which lost heat faster, the land or the water? How do you think this affects the night temperatures of a town located at the edge of a large body of water? Were water and land both heated to the same temperature when you took the first reading?

TESTING THE WATERS
An investigation of solar heat distribution

Does the water in a small lake evenly distribute the sun's heat energy, so that the temperature of surface water is the same as the water several feet down? Hypothesize, or guess, which you think will be warmer, the water on the surface of a small lake, the water several feet down, or the water at the edge, along the shoreline.

It is easy to take a lake's temperature near the shoreline. All you need to do is place a thermometer into the water at the edge of the lake. To read the real temperature of the lake's surface and some distance below is a little harder.

YOU NEED

- thumbtacks or nails
- 6-inch-long (15 cm) piece of wood (1 x 2 or 2 x 4)
- fishing rod and reel
- string
- a thermometer
- a thermometer that floats (an aquarium thermometer)
- metal washers
- a lake or pond
- pencil and paper
- an adult

The thermometers should be placed near the middle of

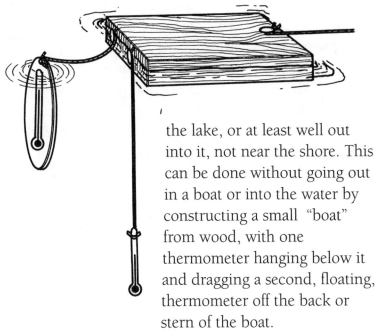

the lake, or at least well out
into it, not near the shore. This
can be done without going out
in a boat or into the water by
constructing a small "boat"
from wood, with one
thermometer hanging below it
and dragging a second, floating,
thermometer off the back or
stern of the boat.

First, find a standard-size piece of wood about 6
inches long (15 cm) to serve as the boat that will tow
the heat-measuring instruments. Tie a long piece of
string to the regular thermometer and fasten the other
end to the boat, using a thumbtack or nail. The
thermometer should hang down at least 2 feet (60 cm)
from the wooden boat. Tie a short piece of string to the
thermometer that floats. Fasten the other end of the

string to the boat also. Floating thermometers are often used in fish tanks and can be found at aquarium supply stores or pet shops.

Now that you are ready, ask an adult to take you and all your equipment to a nearby pond or lake and stand by to help with this experiment.

First, you need to find out the temperature of the lake at the shoreline. Bending carefully on the shore of the lake, place the thermometer in the water. Wait about three minutes to give the thermometer a chance to reach the proper temperature. Then remove the thermometer and immediately read and write down the temperature that it is registering.

To get the wooden boat out into the lake, lay it securely on the beach at the edge of the water or have your helper hold it there. Fasten the end of a fishing line from a fishing pole to the boat with a thumbtack. Slowly let out the line as you walk around the shore of the lake. (A helper could hold the boat and release it at your signal, when you get into position on the other side.) Carefully reel in the fishing line, dragging the boat and its "instruments" to a spot near the middle of the lake (or at least to a spot where the water is deep). Wait three or four minutes to allow the thermometers to properly change temperature. Then, reel the boat in as quickly as you can, and read the thermometers. The thermometers must be read before they begin to change temperature. The shoreline where you stand to reel in the boat should have a sandy or a soft bottom, because quickly dragging the thermometers over a rocky bottom might break them.

Which thermometer read the warmest temperature? Was your hypothesis correct?

POWER RANGER

Measuring home electrical energy usage

Electric power is measured in units called "watts." You've probably heard someone ask for a 40-, 60- or 100-watt bulb to change a burned-out light bulb. The number indicates how much electric power the bulb uses to reach full brightness. (Which do you think is brighter, a 60-watt or 100-watt bulb? Which uses more power?)

Every month, the electric company bills people for the amount of electricity they used. Electrical usage is measured in "kilowatt-hours." One kilowatt equals 1,000 watts. One kilowatt-hour is 1,000 watts of electric power being used for one hour. It takes one kilowatt-hour of energy to operate ten 100-watt light bulbs for one hour.

How much electrical energy does your home use in one day? Find the electric meter for your home (usually outside, where someone from the power company can find and read it easily). The meter's face

has dials on it, marked
with numbers. The
dials, reading from right
to left, show the ones,
tens, hundreds, and then
the thousands places. When a
dial needle is between two numbers, it
is the lower number that is read (a needle
between 2 and 3 is read as 2).

Before school, read the numbers on the
electric meter dials and write them down. The next
day, at the same time, read the numbers again and
record them. Subtract the second day's numbers from
the first reading to find out how many kilowatt-hours
of energy your home used over that 24-hour period.

Make a list of the appliances in your home that use
electricity from the power company. Saving energy is
good for the environment and will save your family
money on the monthly electric bill too. How do you
think you can use the appliances on your list more
wisely to save energy in your home?

THE RIGHT STUFF

Seeds store enough food energy for germination

If you were traveling on foot for a long period of time, you would carry a backpack. In the backpack you would store all the food you need to get you to the next camp, where you could replenish your food supply.

In the same way, seeds store just enough energy to be able to grow a root and a leaf. Once a seed has formed a root to gather water and nutrients from the soil and a leaf to collect sunlight, it can begin to make food on its own. The process of a plant making its food by gathering the light energy from the sun is called photosynthesis. Also needed in the process are carbon dioxide, water, chlorophyll (which gives leaves their

YOU NEED

- 3 different types of seeds (vegetable or flower)
- potting soil
- a dark place (a closet or basement)
- water
- 3 small containers (plastic drinking cups, etc.)
- masking tape
- pencil and paper

green color) and very small amounts of minerals. The time from when a seed begins to sprout a root and a leaf (using its own stored energy) until it is able to make food on its own is called germination.

At your local hardware store or garden center, buy 3 different packages of seeds and a small bag of potting soil. The seeds can be flower or vegetable seeds.

Now, prove that seeds store enough energy to germinate, but then need sunlight to make food in order to continue to grow and live.

Fill 3 small containers with potting soil. Plastic drinking cups or short drinking glasses work well. Get

3 different kinds of seeds; they can be flower seeds
(morning glory, marigold, etc.) or vegetable seeds
(radish, lima bean, watercress, etc.). Place a piece of
masking tape on the side of each container and on
each write the name of the seeds you are going to plant
in the container. Then put 5 of each kind of seed in
their proper container (5 seeds are planted in case
some do not germinate). Push the seeds about ½ inch
(about 13 cm) down into the soil.

Place the containers in a dark place (a closet, for
example) that is at least as warm as room temperature

all the time. Water the seeds every day. Keep the soil moist, but not heavily soaked.

Keep a written log of your observations each day. Write down the date and what you see in each container.

Once leaves appear, the stored food energy in the seeds is just about gone, and the plants are ready to begin making their own food. If the plants are kept in the dark and don't get any light to make new food, how long does it take for them to use up their stored energy and begin to die?

If you keep close watch and provide water and light at the right time, the young plants that start withering for lack of food may begin to start getting food from the soil. If not, plant new seeds and give them TLC (tender, loving care), and you may get to see them thrive.

IN THE PINK
Home insulation keeps heat in and cold out

When the cold winter winds blow, we need to keep the heat energy from our home's heater inside. Builders use a material called "insulation" to keep warm air inside in the winter and cool, air-conditioned air inside in the summer. Insulation looks like thick blankets of cotton candy, usually pink or yellow in color. It is placed inside walls, ceilings, and sometimes under the floor.

YOU NEED
• 2 shoe boxes
• adhesive tape
• clear plastic food wrap
• scissors
• 2 thermometers
• modeling clay
• a sunny window
• sheets of Styrofoam (about ½ inch or 1 cm thick)
• paper and pencil

Home insulation is given an "R-rating," which stands for how good a job that particular kind of insulation does. The higher the R-rating number, the better the insulation is at keeping the temperature on one side of the insulation from changing the temperature on the other.

How do we know that insulating materials do what they are supposed to? Let's prove it.

Remove the lid from a shoe box, or tear the flaps off of another box about the size of a shoe box. Stand the box upright on one end, so it is tall. With scissors, carefully cut a window in the "front" of the box, as shown. The window should be in the upper ⅓ of the box. Then cover the window by taping a piece of clear plastic food wrap over it. Do the same to a second shoe box.

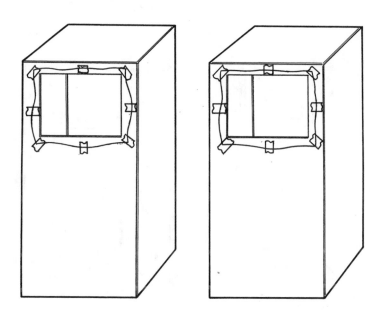

Turn the boxes around to work on the open "back" side.

Styrofoam is a light, usually white material. It is used for many things, including disposable coffee cups and for packing, so that appliances such as TV sets and

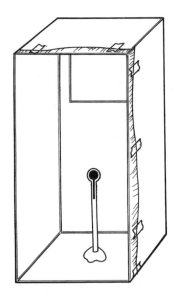

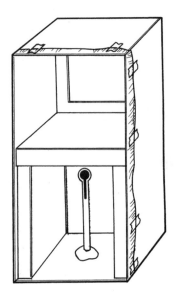

microwave ovens are not damaged in shipping). It is also an insulating material, inexpensive, which is available in many shapes and sizes at hobby shops and craft stores.

Using adhesive tape or glue, line the bottom half of one shoe box with sections of Styrofoam, covering the three walls and making a "roof."

Inside each shoe box, place a small mound of modeling clay on the bottom. Turn a thermometer upside down and stick it in the clay. Do the same in the other box. The thermometers' bulbs will measure the air temperature inside the boxes.

Cover the open backs of each box with a piece of clear plastic food wrap. Use adhesive tape to make the wrap fit tight.

Place both boxes in a sunny window, with the open "back" side of the boxes facing away from the sunlight. Be sure that the sunlight is not shining directly on the thermometer through the window in the uninsulated box.

Over a period of one or two hours, take readings and write down the temperature showing on both thermometers every five minutes. Does the air inside the insulated part of the shoe box stay cooler longer than the air in the uninsulated box?

MARBLE ROLL

Converting kinetic energy into potential energy

Energy can be transferred from one object to another. "Kinetic energy" is the energy of an object in motion. If an object is moving and it hits another object, its energy is transferred, or handed over to the other object. On a pool table, players hit a ball with a cue stick, giving kinetic (movement) energy to the ball. When the ball rolls into another ball, all or part of this energy is given to the second ball, and the second ball begins to roll. Even though the first ball may stop, the force of kinetic energy continues on in the second ball.

> ### YOU NEED
> - marbles
> - 2 rulers
> - 2 books
> - a rug, or carpeted floor
> - piece of paper

Near a rug, or on a floor that has carpeting on it, lay two rulers next to each other, leaving a small space between them. Lay marbles back to back all along the space between the two rulers. Be sure the rulers are close enough together so that the marbles are not touching the floor—but are being held up by the rulers.

At one end of the rulers, open a book to about the middle. Make a ramp out of the open book by propping up the end opposite the rulers with another book or two.

Hold a marble at the top of the ramp. Let go. The force of gravity will give the marble motion energy. When the marble hits the first marble on the rulers, it hands its energy over to that marble. Being up against another marble and unable to move, that marble transfers its energy to the next marble. The energy continues to move from one marble to the next until the last marble is reached. When the energy is given to the last marble, it begins to roll, because there isn't

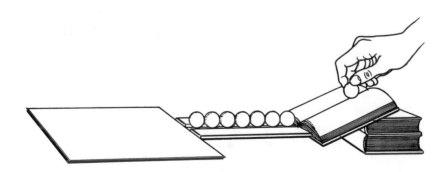

anything blocking it. Is there enough energy to knock the last marble off the rulers? How far does it roll? The rug or carpet offers friction to the marble, and helps slow it down. With a tiny piece of paper, mark the spot where the marble comes to rest, or stops.

Now, remove a few of the marbles so there are gaps between some of the marbles as shown in the illustration below. Roll the marble down the ramp again. Does the end marble roll as far? If not, why not? Do you think it may be because some of the energy was lost before getting to the last marble?

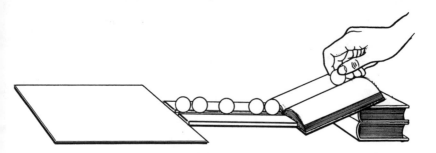

FROLICKING IN THE WAVE
How some energies move

Energy can travel in the form of a wave. You are familiar with rolling waves in the ocean. Other types of energy waves, such as sound waves and radio waves, would look similar to ocean waves, if we could see them.

YOU NEED
- jump rope
- a fence
- a friend
- eye dropper
- round cake pan
- water

A wave has a "crest" or peak, the highest part of the wave, and it has a "trough" or valley, the lowest part. The length from crest to crest (or trough to trough) is called the "wavelength." The wavelength of a tsunami (a tidal wave) can be 100

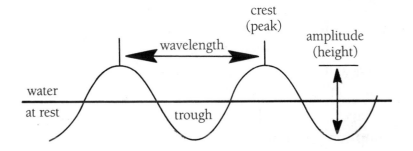

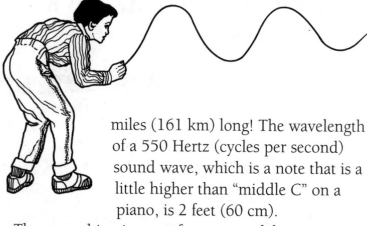

miles (161 km) long! The wavelength of a 550 Hertz (cycles per second) sound wave, which is a note that is a little higher than "middle C" on a piano, is 2 feet (60 cm).

The same thing is true of waves in a lake or an ocean. When you see a rolling wave, it might look like the water is rolling, but very little water is actually moving. The water is only rising and falling as the force of energy passes through it. A boat on the water will bob up and down as the wave energy moves under it. You can demonstrate this by placing a cork or a toothpick in a tub of water and dropping in a stone. The waves created by the stone entering the water will ripple out, and will push the cork or toothpick up and down.

Have a friend hold lightly onto one end of a long jump rope as you hold the other end. Quickly whip your end up and down. You will see the rope take the shape of a wave, which travels down the rope toward the other end. If your friend is holding the rope, when the energy gets to his end, it will yank the rope out of his hand.

Now tie one end of the jump rope to a fence. Quickly whip your end up and down, again and again, and set up a wave pattern. If you want to measure the "amplitude" or height of the wave, have a friend stand alongside the rope and hold up a measuring stick, and watch where the lowest point (the trough) and the highest point (the crest) fall.

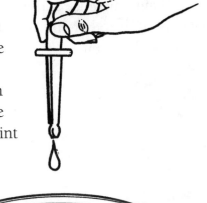

Fill a round cake pan half full of water. Fill an eye dropper with water. You can find the exact center of the pan by squeezing a drop of water into the pan and causing rippling waves. If you squeeze a drop of water perfectly in the center, the waves will ripple out to the edges of the pan, then back again, and meet exactly at the center point. If the waves didn't meet back together, wait until the water calms; then keep trying it until you find the exact center.

Sound wave energy can also travel through other materials besides air. Put your ear on the metal pipe of a chain link fence and have a friend tap a nail on the pipe at the other end. You will hear the sound travel through the pipe.

LITTLE SIR ECHO
Making use of sound waves to measure distance

The energy of a sound wave travels through the air at about 1100 feet, or 335 meters, per second. Energy waves can travel through a medium (air, water,

YOU NEED
- a large building
- tape measure
- friends

metal, etc.) but the molecules of the medium don't actually travel forward with the wave.

As a rolling wave moves through the ocean, the wave energy moves forward, but the actual molecules of water only move up and down with the wave's crest and trough. That is why a boat on the water will bob up and down, but not move forward, as a rolling wave goes by. This concept can also be shown by tying a piece of ribbon onto the middle of a jump rope, then tying one end of the rope to a fence, and moving the other end up and down, setting up waves as in the previous experiment. The ribbon will go up and down, but not forward in the direction of the wave.

The energy of sound waves also travels in the same way. Molecules in the air bump into each other and push the wave along, but the actual molecules travel very little.

Since we know that sound travels at about 1100 feet per second, we can use it to determine distance. You are probably familiar with counting "one one

thousand, two one thousand, three one thousand," to count the seconds between seeing a flash of lightning and hearing the rumble of thunder. Since 1100 feet is about one-fifth of a mile, a gap of one second between seeing a lightning flash and hearing its thunder indicates that the lighting is about one-fifth of a mile away.

Have you ever gone into an empty room and heard your voice echo off the walls? When your ears hear two sounds that are about one-tenth second or longer apart, your brain interprets those as two distinct sounds...an echo, if the two sounds are the same (a loud yell, for example). Traveling at 1100 feet per second, it takes a sound wave about one-tenth of a second to go 110 feet. If you yell at a large wall, you will hear an echo if the sound travels 110 feet or more to get back to you. That would mean you are 55 feet from the wall, because the sound would have to travel to the wall and back to you ($55 \times 2 = 110$).

Find a building that has a broad wall, perhaps your school building. Face the wall and stand about 30 feet away from it. Yell loudly at the wall. Take a step backward and yell again. Continue to move back until you hear your voice echo. Then, using a tape measure, measure the distance from the wall to where you stood when you first heard an echo. Is the distance about 55 feet?

Have your friends try it. Record the distance where each friend first hears his echo. Are the distances close to 55 feet? Add the distances together and divide by the number of friends to find the average distance. How close is the average distance to 55 feet?

SUN, YOU'RE TOO MUCH!

Taming solar energy the natural way

Heat energy from the sun is usually thought of in a good way because of the many benefits. But there are times when this heat is unwanted. Have you ever walked on a beach when the sun made the

YOU NEED
- thermometer
- a sunny day
- a large shade tree
- pencil and paper

sand so hot that you had to run or put shoes on your feet? Or gotten into a car "baking" in the hot summer sun with the windows closed up?

In winter, bright warm sunshine streaming through the windows of your home helps keep you comfortable inside. But in summer, this extra heat causes fans and air conditioners to work even harder, as they try to cool down the house.

Trees can provide natural shading for homes. By planting trees along the side of a house where the hot summer sun beats, a home can be kept cooler naturally, and save electrical energy.

Let's prove that trees lower the temperature of the air in their shade. On a hot sunny day, find a large tree and stand in its shade. There, hold a thermometer out at shoulder height, being careful not to touch the bulb. Wait several minutes for the temperature to settle, then read the thermometer and write down the temperature.

Now, stand in the open, in the sun away from any shade. Again, hold out the thermometer and wait for the temperature to settle. Then read and write down the temperature.

Which location had the lower temperature?

THE GREEN SCREEN

Trees as natural wind-energy protection

Protecting a home from strong, cold winter winds would certainly help to keep heat energy inside the house and lower energy costs.

YOU NEED
- art and construction supplies

Farms often have large open areas in which acres of crops are planted. To slow strong winds down and protect their crops and topsoil from wind damage, and their homes from cold, farmers often plant trees in a row to grow tall and cut the prevailing winds. Long ago, during long, cold winters, such windbreaks were especially important to the families living in farmhouses, when insulating materials and efficient means of heating weren't as good as today.

Using art and construction supplies or scale models (the kind used for train sets), construct a model of a farm, showing its open, crop-filled fields; the farmhouse; and where rows of trees would be placed to act as windbreaks. Research weather maps in the area of your "farm" for wind direction.

Again, using art and construction supplies or models, construct a model of a home or apartment dwelling in a city. A city home may only have exposure to the wind in the front and back, if it is attached or very close to other buildings on both its sides. Other people's homes or apartments will act as windbreaks for the person's home in the middle.

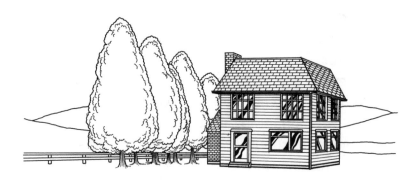

NOW YOU HEAR IT...
Tracking radio frequency direction

At radio and television stations, powerful transmitters send radio waves through the antenna and into the air, on their way to your radio receiver

or TV set. Can you use a radio as a direction finder in order to locate the station from which a radio wave is coming?

Take an inexpensive AM pocket radio and place it on the table. Tune it to a station that is coming in strong and clear. Fold a piece of aluminum foil in the shape of a cave or pocket and place it around the top, bottom, and three sides of the radio, leaving only the front of the radio showing in the opening. Place the radio back in position on the table.

Is the station still coming in as strong? Slowly turn your radio, along with its aluminum shield, exposing the open front side to different directions. Is the station still coming in as strong? Continue to turn it around until you have made a complete circle. If the station is strongest when facing one particular direction, then that is the direction that radio waves are coming from.

Do the experiment with several other radio stations. Are the signals coming from the same or different directions? List the stations, by their ID letters or the numbers where they are on the radio dial, and write down the direction of their signal transmitter.

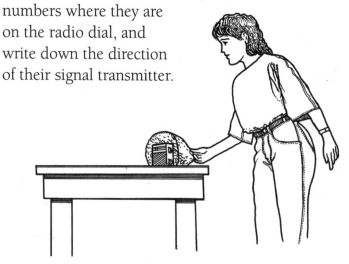

ONCE, TWICE, AGAIN
Reusing products to save energy

It takes energy for a factory to make a product. If that product can be used twice, or used to do more than one job, that reduces the need to make more of the product—saving energy.

YOU NEED
- various materials from around the house (plastic soda bottles, paper grocery bags, plastic bags, plastic margarine tubs, etc.)
- pencil and paper

What items around your home or school can be used more than once, or used in a different way after their first use? Plastic products are often handy to keep for other uses. Responses to a recent consumer poll suggest that 87 percent of Americans had reused a plastic product over the previous six-month period.

SOME ADDED USES FOR
MANUFACTURED PRODUCTS

Plastic soda bottles

1. Make into bird feeders.

2. To store cold drinking water in refrigerator.

3. Use as a terrarium.

4. Experiments.

5. A penny bank.

6. Pocket money— return bottles to store for refund.

Can you think of other uses for bottles, and for the other products listed here?

Paper grocery bags

1. Bring to grocery, when out shopping, to pack order.

2. As covers for school textbooks.

3. To carry things, such as gifts, when visiting relatives.

4. To wrap packages for mailing.

5. For arts and crafts projects, such as masks for plays or Halloween.

6. Spread out to protect work area.

7. Place them in your community's recycling container.

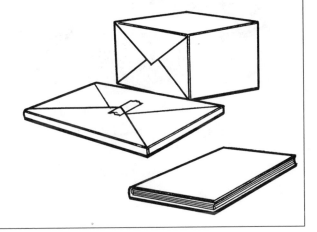

Plastic margarine tubs

1. To store screws or small items.

2. To store food leftovers.

3. As scoop—for dog food, animal feed, fertilizer.

4. To catch water under small flower pots.

5. With puncture holes in bottom to sprinkle water on plants.

6. As beach toys, to hold sand and water, or use as molds for sand castles.

Plastic grocery bags

1. Line small bedroom wastebaskets.

2. To carry school lunches that may be soggy or might leak.

3. To put carried books or packages in when it starts to rain.

4. When packing a wet bathing suit.

5. To hold wet laundry to hang or dry.

6. To keep feet or shoes dry in a rainstorm.

7. To collect recyclables for pickup.

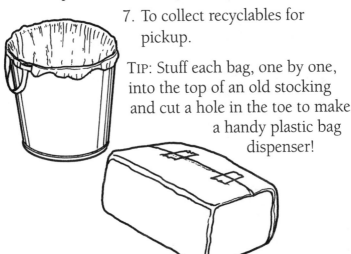

TIP: Stuff each bag, one by one, into the top of an old stocking and cut a hole in the toe to make a handy plastic bag dispenser!

NUCLEAR DOMINOS
Demonstrating a chain reaction

One type of energy is released when the nuclei of atoms are either combined (fusion) or split apart (fission). The energy, called "nuclear energy," is

released in the form of heat, light, or some other type of radiation. Nuclear energy is used to make electricity by heating water for steam that then drives giant turbine generators.

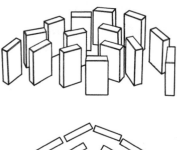

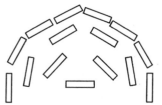

In the process of nuclear fission, the splitting apart of an atom causes a chain reaction. A radioactive element, such as Uranium-235, is used in the chain reaction. The first fission creates two new neutrons. Each of these neutrons strikes at least two other neutrons,

which strike even more, and
a chain reaction takes place
which continues to grow.

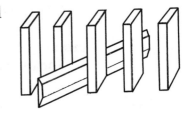

To demonstrate the
concept of nuclear fission,
stand dominos in the pattern
shown, where one domino is set to start a chain
reaction. As it falls, the domino hits two dominos,
which each hit two more. The number of dominos hit
and falling with each row grows quickly. This chain
reaction of falling dominos can be demonstrated easily
using only 4 or 5 rows; with additional rows the
arranged dominos become so packed together that
they get in each other's way.

In a nuclear reactor where nuclear fission is taking
place, the chain reaction can be slowed down or even
stopped by inserting rods made of cadmium or boron.
These dampening rods absorb neutrons and slow the
process down. Demonstrate this by placing a ruler
between two dominos in a row, then push the first
domino to start the chain reaction. The reaction stops
when it reaches the ruler, which acts like the rods in a
nuclear reactor.

BRIGHT HEAT

Unwanted heat energy from incandescent light

The most common types of home lighting are incandescent and fluorescent bulbs. In both, electrical energy is turned into light energy.

In an incandescent light bulb, electricity passes through a small wire, called a filament, which glows brightly. In a fluorescent light bulb (usually a long straight or circular tube), the inside of the bulb is filled with a gas. The inside glass of the bulb is coated with materials called phosphors. When electricity is passed through a heating element in the bulb, the gas gives off rays that cause the phosphors to fluoresce (glow).

All we really want from a light bulb is light. However, in the changing of electrical energy into light energy, there is energy loss. Some of that energy is given off as heat, especially unwelcome during hot weather! Which type of light bulb is more energy-efficient (gives off less heat to the surrounding air)?

Find a lamp in your home that uses an incandescent bulb; most lamps do. Find a fluorescent lamp. In the home, fluorescent bulbs are often used in kitchens, bathrooms, garages, and workshop areas. Remember, when working around hot light bulbs and electrical fixtures, to have an adult with you. A light bulb may stay quite hot even after it has been turned off, and it can cause a burn—so be careful. Also, do not take bulbs out of their sockets.

Remove the shade from a lamp that has an incandescent bulb. Hold the bulb of a thermometer about one inch (2.5 cm) away from the lit bulb for about three minutes. Record the temperature.

Now hold the thermometer bulb the same distance from a lit fluorescent bulb for three minutes. Record the temperature. Which bulb is more light-efficient?

UNEQUAL ENERGY

Finding the distribution of heat energy in a room

Is the temperature in a room the same everywhere in the room? You might think that it is. But as you do this project, you may be surprised to discover that the air temperature is different in different parts of the room—

YOU NEED

- a large room in your house
- thermometer
- a clock or watch
- paper and pencil
- an adult

the heat energy in the room is not equally distributed.

Pick about ten different locations in a large room in your home. Some locations should be high up, some low near the floor, some on an "inside" wall (a wall with another room behind it), and some on an "outside" wall (with the outdoors on the other side). High locations can be above an inside door (over the open doorway or, with the door ajar, the thermometer lying on top) or on top of a picture frame on a wall. Ask an adult to help you place thermometers in the highest places. One of the locations should be by a window, another by an electric light switch on an outside wall.

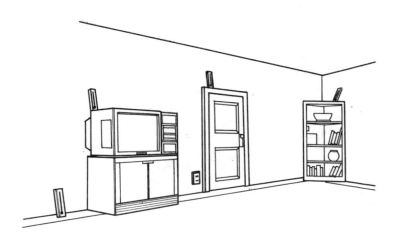

Make a chart with two columns and at the head of the first column write "locations"; then list the locations you've selected. At the top of the second column write "temperature."

Put the thermometer at your first location. Wait about three minutes to give the thermometer time to adjust and indicate the correct temperature. Write down the temperature reading for that location. Repeat this procedure for all of the locations in your room.

Compare the temperatures you have taken around the room. Which location is the warmest? Which is the coolest? Why do you think the temperature may be different at each location? Look up the word "convection" in the dictionary. Do you think poor insulation by a light switch on an outside wall or by a window might be affecting the air temperature at that location?

BAND AT TENSION
Measuring potential energy in a stretched elastic band

When you pull the elastic band of a slingshot back as far as it will go and hold it, the elastic band has potential (stored) energy, ready to do work. When you let go, that potential energy is released to do work. This energy in motion is called kinetic energy.

How can we show that the more a rubber band is stretched, the more potential energy it has (and the more kinetic energy is released when you let go of the rubber band)?

Let's construct a paper towel tube "cannon" and use a Ping-Pong ball as a cannonball to measure kinetic energy when it is shot out of the cannon. Don't

YOU NEED
- Ping-Pong ball
- masking tape
- empty paper towel roll
- thick book
- measuring tape
- rubber band
- board, about 3–4 inches (8–10 cm) wide and 1 foot (30 cm) long
- ruler
- 2 nails
- hammer
- paper and pencil

use anything heavier as the cannonball, because flying objects can be dangerous. A Ping-Pong ball is safe to use.

Take two pieces of masking tape about 4 inches (10 cm) long. Lay them on top of each other, with sticky sides touching.

Fold the masking tape over one end of an empty paper towel roll. Position it so that it covers only a part of the paper towel roll opening as shown. Use another piece of masking tape to hold it onto the roll. This tape will act as a "stopper" by making the opening just small enough to keep the ball from falling through, but will let it stick out of the bottom a little. Lay the roll aside for now.

Using masking tape, tape a ruler to a piece of wood about 3 or 4 inches (8–10 cm) wide by 1 foot (30 cm) long. Let about 5 to 6 inches (13–15 cm) of the ruler hang over one end of the board.

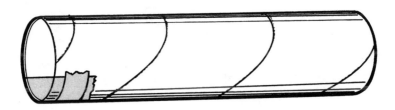

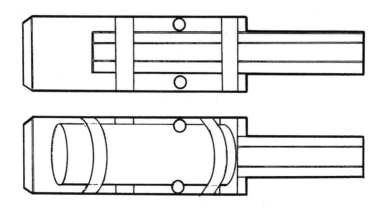

Hammer two nails part of the way into the board, one on each side near the outside edges, about 3 inches (7.5 cm) from the end of the board where the ruler is. Leave the nails sticking up slightly.

Lay the paper towel roll on top of the ruler and board assembly, between the nails, and fix it in place with masking tape. Set your "cannon" on the floor.

Elevate the front end of the cannon by placing a book under the end opposite the ruler. Put a rubber band across the two nails, stretching it around the bottom of the paper towel tube. Drop your "cannonball" into the tube so it comes to rest at the bottom of the tube, on the masking-tape stopper and the rubber band.

Using the scale on the ruler, pull the rubber band back 1 inch (2.5 cm) and let go. The rubber band will hit the ball and shoot it out of the cannon. Watch where the ball first touches the floor. Use a tape measure to find the distance the ball traveled out of the cannon. Make a chart to record how far back the rubber band is pulled and how far the ball travels through the air.

Next, pull the rubber band back 1½ inches (4 cm). Measure and record the distance the ball travels. Repeat this procedure, ½ inch (1 cm) at a time, until the rubber band has been stretched back as far as it can go.

Look at your chart of data. Does the ball travel farther if the rubber band is pulled farther back? Is there a mathematical relationship (a number pattern)

between the distance the rubber band is pulled back and the distance the ball goes? For example, does the ball go 1 foot (30 cm) farther for each inch (2.5 cm) the rubber band is pulled back?

HEAT WAVE

Discovering how microwaves generate heat

Microwaves are a kind of radio frequency energy (electromagnetic waves). Their frequency (the number of times the wave vibrates each second) is much higher than most other types of radio and TV waves.

Microwaves are used for telephone and satellite communications, and for fast cooking. When microwaves pass through food, they cause the molecules in the food to move back and forth very rapidly. This generates heat. Have you ever rubbed your hands together rapidly to warm them? A microwave oven works in a similar way. Microwaves vibrate the molecules of water, sugar and fat in food, but pass right through glass, pottery, paper, wood and plastic. That is why, although food cooks in a microwave oven, the dish doesn't get hot—except for some heat

YOU NEED

- use of a microwave oven
- 2 thermometers
- coffee or tea cup (must be "microwave safe")
- water
- an adult

transfer from the food. Metal blocks microwaves, so should never be used in a microwave oven.

Can you prove that a microwave oven does not cook by making the air in the oven hot, like a traditional oven does, but by heating up the food from the inside?

Ask an adult to help, and fill a cup with water. Put it inside a microwave oven and heat it for 60 seconds. Be sure the cup is "microwave safe," that is, made of plastic, glass or pottery without any metal in it or metallic decorations on it.

When the time is up, take the cup out of the oven. Lay a thermometer inside the oven and close the door (but DO NOT turn the oven on). Stick a thermometer into the cup of water. After about two minutes, take the thermometer out of the oven and compare it to the one from the cup of water. Does the thermometer that was in the water read a higher temperature than the one placed in the oven?

STATIC INTERFERENCE

Detecting sources of stray radio frequency energy

One kind of energy that we can't see or hear is called radio frequency energy. Radio frequency energy is formed by electromagnetic waves that travel through the air. This energy allows us to communicate with one another, and is used to bring TV pictures and sound to our homes from stations far away. Two-way radios let people talk to each other from remote places (you may even own a pair of "walkie talkies"). Cordless telephones allow people to talk on the phone while walking around the house or going outside without being restricted by a wire connecting the handset to the telephone base. "Cellular phones," both hand-held and car phones, permit communications with people who

YOU NEED

- electric shaver
- cordless telephone
- fluorescent light
- electric blanket
- electric hair dryer
- AM radio
- personal computer
- television
- paper and pencil

are not near a regular telephone. All of these types of communication are possible because of an invisible kind of energy called radio frequency energy.

Some things may give off radio frequency energy, even when we don't want them to. Things that have electric motors often produce radio frequency energy

when they are working. This energy may be unwanted, since it can interfere with other things that use radio frequencies, such as TVs, radios, and cordless telephones.

What things around your home do you think might be radiating (giving off) radio frequency energy? How about an electric shaver, a fluorescent light, an electric blanket, an electric hair dryer, an AM radio, a personal computer, or a television?

Since we cannot see or hear radio frequency energy, we will use a radio as a detector to help us find things around the house that are producing radio frequency energy.

Tune a portable AM radio to a spot on the dial where no station is heard. Bring the radio close to each of these objects:

- an electric shaver
- a fluorescent light (often found in the kitchen or bathroom)
- an electric blanket
- an electric hair dryer
- a personal computer
- a television (turn the volume down on the TV set)

Write down your observations about each appliance. Did you hear a sound in the radio? If so, describe the sound; was it a crackling sound or a humming sound?

Call a friend on a cordless telephone. Ask your friend to be quiet and listen. Then hold your phone close to each of the appliances listed above, asking your friend each time what she or he hears, if anything. If your friend has a cordless phone, it's your turn to listen. What do you hear?

What other things around your home or school can you check out as radio frequency producers? What do you think—which makes the better radio frequency detector, an AM radio, or a cordless telephone?

CELL MAGIC

Changing light energy to electrical energy

The photovoltaic cell, better known as a "solar cell," is a device that turns light directly into electricity. Solar cells are expensive to manufacture, so they are only used when there is no other easier way to get electricity, such as at a remote weather station or in an Earth-orbiting satellite.

A single solar cell does not generate very much electricity, but solar cells can be connected together "in series" and their individual voltages added together. Connecting cells in series means hooking

YOU NEED

- 3 hobby solar cells (delivering about ½ volt each)
- small DC hobby motor (requiring 1½ to 3 volts direct current)
- insulated jumper leads with alligator clips on each end
- a sunny window with a shade, blinds, or curtain
- several small wooden blocks

the positive (+) terminal of one cell to the negative (–) terminal of the next. Flashlights have their batteries connected in series, with the negative terminal of one

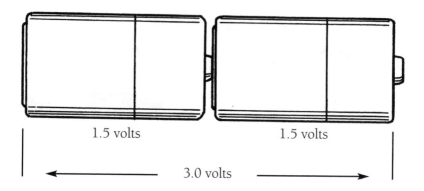

1.5 volts   1.5 volts

← ———————— 3.0 volts ———————— →

battery touching the positive terminal on the next. When batteries are connected in series, the total voltage available across all of them is the sum of the individual battery voltages as shown. Imagine how much more power you could exert on a rope if the strength of three of your friends were also helping you to pull it.

Using insulated jumper leads with alligator clips on each end, connect the positive and negative terminals of a solar cell to a small 1.5-volt hobby motor. Set the arrangement in a sunny place. Use wood blocks behind the cells to tilt them so that they face the sun. You could also use spring-type clothespins clipped onto the sides near the bottoms of the cells to stand

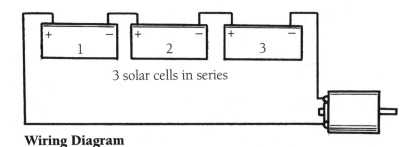

Wiring Diagram

them upright. Watch how fast the motor spins. Next, add two more solar cells to the circuit, placing them in series.

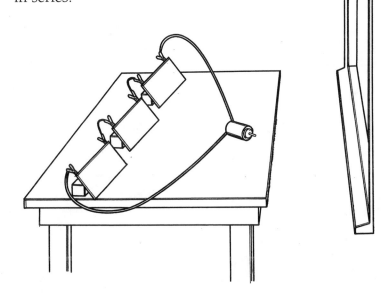

Look at and listen to the motor. Is it spinning faster now that it is getting more voltage? What happens to the motor's speed on a cloudy day?

Create your own "cloudy day" by closing blinds or curtains partway, then all the way. Do you think the voltage produced by the solar cells is less on cloudy days? How do you think that affects the location of where solar cells work best?

PENNY SHOOT

Newton's law and the transfer of energy

Sir Isaac Newton, (1642–1727), an early English scientist, formulated the physics laws of motion. His first law of motion states that "an object at rest tends to stay at rest, and an object in motion tends to stay in motion." A famous trick demonstrates this law. A playing card is placed over a drinking cup and a coin is laid on top of the card, directly over the mouth of the cup. Then the edge of the

card is given a sharp tap with a finger or pencil. The blow knocks the card off the cup, but the coin on the card, being at rest, stays in position. Then gravity pulls down on the coin and it drops into the cup.

Let's build a device that will not only demonstrate this law of Newton's, but also show transfer of energy (energy from one object being handed off to another object).

Cut a strip of smooth paper to fit on the wooden board. Toward one end of the board, hammer two small nails, spaced about 3 inches (7.5 cm) apart, partway into the wood. The nails should be sticking up out of the wood about an inch (2.5 cm), looking like goal posts at a football field.

Stretch a small rubber band between the two "goal post" nails. Cut a small strip of thin cardboard, about an inch (2.5 cm) wide by two inches (5 cm) long. Fold it in half around the rubber band (in the middle) and staple the cardboard ends together. Push the rubber band down on the goal posts, so it rests almost against the wood.

About one inch (2.5 cm) in front of the goal posts, place one of the medium coins face up. Then stack four more of the coins on top, but with tails up.

Grab the stapled piece of folded cardboard between your thumb and index finger and pull back. A

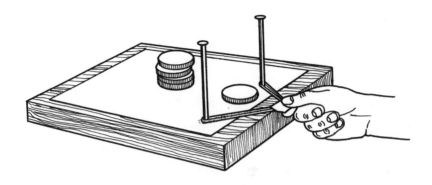

stretched rubber band is said to have "potential energy," energy that is stored up and ready to do work.

While the rubber band is stretched, place the smaller coin between the goal posts. Release the cardboard so that it strikes this coin and shoots it toward the stack. The idea is to knock the bottom coin out from under, leaving the other four stacked coins at rest (although they will drop straight down due to gravity).

You may have to try this several times. Your aim may be off, and sometimes the smaller coin may fly slightly upward and miss hitting the bottom nickel. If a coin does move from the stack, it may happen too fast for you to see. To be sure it was the *bottom* coin that really was knocked out, see if the coin that was moved has heads or tails up. If it's heads, then you successfully shot out the bottom nickel.

This project also shows two examples of the transfer of energy. When the stretched rubber band (potential energy) is released (kinetic energy), energy from the rubber band is transferred to the smaller coin, giving it motion. Energy is then transferred again when this coin hits the stacked coins. The force must be great enough for this struck coin to overcome the friction of the coins on top of it—and the surface under it, which is why we placed a piece of smooth paper under the stack.

The "momentum" of the moving coin will determine just how far it will travel after it has been shot out from under the pile. Remember, "objects in motion tend to stay in motion," so once the nickel is moving, it will *naturally* try to keep going; friction eventually slows it down enough to make it stop.

PHYSICS

PHYSICS

Physics

Welcome to the fascinating world of physics! This book explores projects in the field of physics. Physics is the science of investigation that tells us the "how" and "why" about nonliving objects. It explains how a refrigerator keeps things cold, why letting the air out of a balloon causes it to fly wildly around the room, and what makes a walkie-talkie work. It tells us why we see a bolt of lightning before we hear its rumbling thunder. Physics helps us unlock the secrets of the physical world around us.

Subjects in physics are numerous, and they include: light, sound, heat, simple machines, forces, magnetism, gravity, friction, acceleration, momentum, time, space, fluidics, pendular motion, wave motion, kinetic and potential energy, work, friction, pressure, weight, conduction, the state of matter (solid, liquid, gas, plasma), electricity, radiation and many more. Physics is one of the most interesting and motivating science topics.

It is important to understand the laws of physics because so many of its principles are found in other science disciplines, such as astronomy, geology, mathematics, health, engineering, electronics, chemistry, aviation, optics, and even the arts. For instance, meteorology, the study of weather, involves many principles that are explained by physics: convection, evaporation, condensation, temperature, precipitation, tidal action by the forces of gravity, temperature, and erosion. Many of these fields (electronics and structural engineering, for example) are really specialized branches of physics.

Physics affects our daily lives. Its principles are at work when we ride a bicycle, wear a pair of glasses, play a computer game, operate a vacuum cleaner, turn on a bedside light, play a music CD or call a friend on the telephone. Physics is at work all around us all of the time.

MAGNETIC WATER
The effect of water on magnetism

Purpose Does water affect a magnetic field?

Overview Sound waves go through both water and air. In fact, they travel farther and faster in water than they do in air. How about magnetism? Does it go through water, too?

Hypothesis Water has no effect on magnetism.

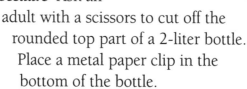

YOU NEED
- magnet
- 2-liter plastic soda bottle
- masking tape
- metal paper clip
- 2 pencils
- string
- water
- scissors
- an adult

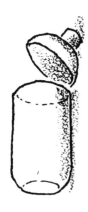

Procedure Ask an adult with a scissors to cut off the rounded top part of a 2-liter bottle. Place a metal paper clip in the bottom of the bottle.

Wrap a strip of masking tape around one end of a six-sided, not round, pencil and then number the sides. Write "1" on the tape on one

side, then turn the pencil and
write "2," and so on. Tie a
piece of string to the middle
of the pencil and secure it
with a piece of masking tape.
Tie the other end of the string
to a magnet. Turn the pencil,
wrapping the string around it,
and set it over the top of the
plastic bottle. Slowly, lower
the magnet into the bottle.
When the magnet is close

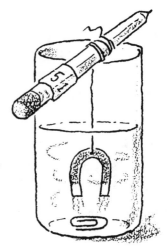

enough and captures the paper clip, stop! Notice the
number on the side of the pencil.

Carefully, lift the magnet straight up without turning
the pencil. Remove the paper clip, and lay it back in
the bottle in the exact same spot. Fill the bottle half
full with water, then slowly lower the magnet into the
bottle. *Be sure not to turn the pencil,* so that the string
length is not changed. The string length, the position
of the paper clip, and the distance from the magnet to
the paper clip are CONSTANTS. The VARIABLE is the

substance between the magnet and the paper clip: air and water.

Does the magnet still attract the paper clip? If so, does it do so from about the same distance above it as it did when the bottle was filled with air instead of water?

Results & Conclusion   Write down the result of your experiment. Come to a conclusion as to whether or not your hypothesis was correct.

Something more   Now test magnetism using salt water, sugar water or ice water.

RUB THE RIGHT WAY
Friction and surfaces

Purpose Compare the friction on a dry surface to one coated with oil.

Overview Friction is the resistance to motion when two things rub together. Friction is often undesirable. It makes machines less efficient where moving parts come in contact with each other. But there are times when friction is helpful. On the road, it's the friction

between a car's tires and the road's surface that allows a driver to keep control of the car. If a road becomes covered with water, snow, ice or spilled oil, the car becomes harder to steer and to stop. This is especially true on a hill.

YOU NEED
• 2 pieces of wood, about 2 feet long (60 cm)
• 2 small plastic butter tubs with lids
• sand
• vegetable oil
• an old rag
• several books
• ruler
• protractor

Hypothesis Hypothesize that if friction becomes less, an object on a slope will need less of an angle for gravity to overcome friction.

Procedure Make a ramp (the slope) using a piece of wood about 2 feet (60 cm) long and 3 to 4 inches (7–10 cm) wide (a 2-by-4 board works well). To raise one end of the ramp, place several books under one end.

Using an old rag, wipe some vegetable oil onto the board, coating and completely covering the surface. This represents spilled oil on a roadway.

Fill two empty plastic butter tubs with an equal amount of sand, and close the lids.

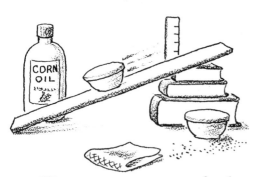

Place one of the filled tubs in the center of the board. By adding more books or pushing them a little farther under the ramp, slowly make the ramp steeper until gravity overcomes the friction between the surfaces and the tub moves. When this happens, stand a ruler alongside the highest point of the ramp. Measure and write down the height of the ramp at that point. Then, using a protractor at the low end, measure the angle, or slope, of the ramp from the table or floor.

Using books and another board, make a ramp with the same slope as the first ramp. Place the tub in the middle of the board. This time, the tub does not move. Slowly raise the slope of the ramp by adding books until the tub finally moves. Measure the angle of the ramp, using the protractor, and see how much steeper

it is compared to the first ramp. The weight here is now the CONSTANT, and the surface friction is the VARIABLE.

What other places can you think of where friction is desirable? Think about walking on patches of ice outside, a newly waxed kitchen floor, or the tile floor in the bathroom when you step out of the shower.

Results & Conclusion   Write down the results of your experiment. Come to a conclusion about your hypothesis.

Something more   Instead of comparing an oil-covered surface to a dry surface, compare a dry surface to one that is covered with ice. Place a piece of wood under the faucet in a sink and run water on it. Then put the wet piece of wood in the freezer and leave it there until the water has turned to ice. Again, find the angle where gravity overcomes resistance and the sand-filled tub moves. Do you think driving a car on an ice-covered road is more dangerous than when the road is dry? Besides driving more slowly, what do people do to help make driving on snow and ice safer?

HOT ROCKS

Heat transfer from one medium to another

Purpose Is there a good way to store solar heat, and release it slowly over time?

Overview Did you ever touch a rock that has been baking in the sun on a warm summer day? Did it feel hot? Rocks can collect and store heat.

Scientists have been working for many years to harness energy from the sun. Solar energy is being used to heat houses. One design uses hollow roof panels so that the sun warms the air inside. A fan blows the warmed air through a pipe to the basement, which is filled with rocks. As the heated air flows over the

YOU NEED

- a scale
- 2 two-liter plastic soda bottles
- small rocks (about the size of small coins)
- large rock (about 3 inches or 8 centimeters in diameter)
- 2 thermometers
- hot water from a faucet
- pencil
- paper
- clock or watch
- masking tape
- scissors
- an adult

rocks, heat is transferred from the air to the rocks, warming them. Then at night, when the collectors no longer gather solar heat, a fan blows air over the rocks, transferring their warmth back to the air. The air is sent to ducts throughout the house to warm each room.

In designing a solar-heated house like this, would it make any difference if huge rocks were used or very small ones? A big rock might have more ability to store heat, but many smaller rocks would have more surface area (they have more sides that would be exposed to the warm air). Find out if a big rock or

many smaller rocks would be better at collecting heat, or if rock size doesn't seem to make much of a difference.

Hypothesis Hypothesize that an equal mass of smaller rocks will absorb heat more quickly than one large rock.

Procedure Have an adult help you by cutting the tops off two 2-liter plastic soda bottles, using a pair of scissors. They should be cut near the top, just at the point where the bottles start to become rounded.

Gather some rocks. One of the rocks should be just large enough to fit inside a 2-liter soda bottle, about 3 inches or 8 centimeters in diameter (across). The other rocks should be small, pebbles about the size of small coins.

Using a scale, find out how much the large rock weighs. Remove it from the scale. Then pile up smaller rocks on the scale until the same weight is reached. The rock mass will then be held CONSTANT, and the size of the rocks is the VARIABLE.

Set all the rocks on a table for an hour or two until you can be sure they are all at room temperature. Do not put them in direct sunlight.

Gather two thermometers. Before we can use them, we must be sure they are calibrated, so that we can use their readings for comparison. (We might have to adjust the reading of one thermometer to correct it so both thermometers read the same temperature.) Leave the two thermometers at room temperature for several minutes, then read the temperature on each one. If one reads higher than the other, put a small piece of masking tape on it and make a note of the difference in temperature. If it is ½ or 1 degree higher than the other, then subtract this much from its reading when comparing the temperature on it to the temperature on the other thermometer.

Have an adult fill each bottle half full of hot water from a sink. Using a thermometer, be sure the water in each bottle is the same temperature. Be careful working around the bottles of very hot water.

Place the large rock in one 2-liter plastic bottle and the smaller rocks into the other bottle. Be careful not to spash the hot water out and on you.

Put a thermometer in each bottle. After a few minutes, record the temperature on each thermometer. Every three minutes, read and record the temperature on the two thermometers. Make up a table, such as the one shown in the illustration, to record your data. Continue to record temperatures until they reach room temperature (about 70 degrees Fahrenheit). Remember to make an adjustment of your readings to calibrate the two thermometers.

Did the water in one bottle cool off faster than the other? If so, then the rock (or rocks) in that bottle collected heat faster.

Results & Conclusion Write down the results of your experiment. Come to a conclusion about your hypothesis.

Something more Which releases heat quickest, one large rock or an equal-mass grouping of smaller rocks? In solar heating for a home, it would be preferable to have heat released slowly over a long period of time, to keep a house warm all through the night until the sun came up again to add heat back into the system.

SMALLER IS STRONGER
Testing tensile strength

Purpose To discover if an object's strength has any relation to its length.

Overview The term tensile strength means how strong something is when it is unsupported; how much tension or pressure it can take before it breaks. Steel has great tensile strength. Is tensile strength affected by length?

YOU NEED

- large metal washers
- 2 paper clips
- string
- 2 hardbound books
- an adult
- 2 long (fireplace) safety matches
- paper and pencil

Hypothesis As an unsupported span decreases in length, it can support more weight.

Procedure Stand two hardback books upright, opening them slightly. Place them about 10 inches (25 cm) apart. *Have an adult* light and blow out long matches, made specially for fireplaces, so they are safe to use. Lay one match across the books. Bend open two metal paper clips so they form an "S," with a hook at the top and bottom of each clip.

Tie a paper clip onto each end of a short piece of string. Hang one paper clip from the middle of the match. Push the hook of the other paper clip through the hole of a large metal washer. This makes it easy to add more washers.

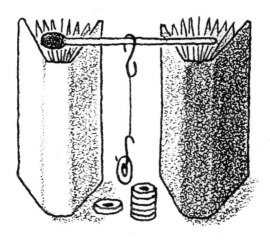

Add washers until the match breaks. Write down how many the match could hold.

Now repeat the experiment, but this time move the books closer together, about half the distance they were. The VARIABLE is the length of the span being

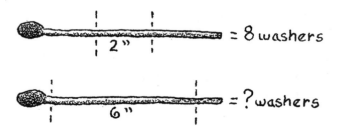

stressed. Will the shorter unsupported span of the match be able to hold more weight without breaking?

Results & Conclusion   Write down the results of your experiment. Come to a conclusion as to whether or not your hypothesis was correct.

Something more   Can you work out (quantify) the relationship between the length of unsupported match in inches and the number of washers needed to break it?

UP TO SPEED

Acceleration in a bottle

Purpose To show changes in rate of speed.

Overview Acceleration is an increase in speed. Physicists define it as a measure of the rate of change of velocity over time. To accelerate means to go faster; to decelerate is to slow down.

Hypothesis It's possible to prove that speed and acceleration are different and measurable by constructing an "accelerometer."

YOU NEED

- 2-liter clear plastic soda bottle
- water
- food coloring
- masking tape
- pen or marker
- tape measure
- an adult with a car

Procedure Partially fill a 2-liter clear plastic soda bottle with water. Add some food coloring so that you will be able to see the water's movement better. Screw the cap on tightly. Place masking tape along one side of the bottle's circumference. Draw a scale on it (millimeters or ¼-inch increments).

Ask an adult with an automatic-shift car to take you for a short drive. (Manual-shift cars could be jerky and uneven during acceleration.) Open the car's glove compartment and use the door as a shelf. Lay your "accelerometer" on it. Fix it there with masking tape or rubber bands. On the tape, mark the water level when the car is not in motion.

When you are ready (remember to fasten your seat belt), have the driver accelerate. Observe how far up the scale the water moves. The faster the car accelerates, the steeper the slope of the water up the side of the bottle. The amount of water in the bottle has remained CONSTANT, but the acceleration of the car has VARIED.

Next, have the driver hold a steady speed, like 40 miles (or 64 kilometers) per hour, on a highway. Is the water level at the same mark as it was when the car was at rest? Even though the car is traveling at quite a good speed, the acceleration is zero.

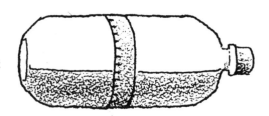

Results & Conclusion Write down the results of your experiment. Come to a conclusion as to whether or not your hypothesis was correct.

Something more What about deceleration, when the car is slowing down and coming to a stop. Can your accelerometer also be used to compare rates of deceleration?

BAD MANNERS

Heat conduction and heat sinking

Purpose Is there a way to make something cool more quickly, like a drink that is too hot?

Overview Metal is a good "conductor" of heat. That means it makes an easy path that heat can travel along. When a metal frying pan is placed on a stove burner, the heat from the burner is conducted (carried) through the bottom of the pan and heats the food inside it.

YOU NEED

- 2 identical containers (coffee mugs or tea cups)
- hot tap water
- 2 thermometers
- masking tape
- spoon
- clock or watch
- paper and pencil

Metal is sometimes used to cool things by conducting heat away from an object. In electronics, transistors and integrated circuits ("chips"), which are found in televisions, stereos and computers, get hot, but heat can damage them. Often metal is made in the shape of fins and attached to transistors and integrated circuits in order to carry the heat away from them. These

cooling fins, called heat
sinks, help transfer
the heat to the
surrounding air
and keep the
transistors and
integrated circuits
cool. Sometimes a small
fan is used to get rid of the
heated air. Does your computer have a fan in it?

Have you ever been served a hot cup of tea or hot
chocolate that was too hot to drink and someone told
you, "Leave the spoon in. It might be bad manners but
it will help cool the drink faster." They are thinking
that, since metal conducts heat, the spoon will carry
some of the heat away from the drink. The spoon is
indeed hot to the touch, so it does conduct heat away
from the drink.

However, since the handle of the spoon is not
designed like a heat sink, the heat in the spoon doesn't
efficiently transfer to the surrounding air, so you may
want to hypothesize that leaving the spoon in the hot
liquid won't make a significant difference.

Hypothesis Hypothesize that leaving a spoon in a hot drink will not make any noticeable difference in its rate of cooling, reducing the temperature faster.

Procedure Gather two thermometers. Before we can use them, we must be sure they are calibrated, that is, we may need to adjust the temperature readings of one thermometer so both thermometers correctly read the same. Leave the two thermometers at room temperature for several minutes, then read the temperature on each one. If one reads higher than the other, put a small piece of masking tape on it and make a note of the difference in temperature. If it is

½ degree or 1 degree higher, then subtract this much from its readings when comparing the temperature on it to the temperature on the other thermometer.

Fill two coffee cups of equal size with equally hot tap water. Be careful working with and around very hot water. Place a thermometer in each cup. Put a metal spoon in one of the cups. After one minute, read the temperatures on the two thermometers and write them down. Every minute, write down the temperatures you read. Be sure to make any adjustment of your numbers to calibrate the two thermometers. Continue to make readings until the water in the two cups reaches room temperature.

Did the water in the cup with the spoon in it cool down faster, or was there not any noticeable difference?

Results & Conclusion Write down the results of your experiment. Come to a conclusion as to whether or not your hypothesis was correct.

Something more Can you find a way that will measurably cool the cup of hot water? Purchase some transistor heat sinks at your local electronics shop and affix them to the cup with rubber bands. Try using a larger spoon, such as a ladle.

ROOM FOR BRIGHTNESS
Reflected light

Purpose Show that a room is better lit when the room's walls are painted in bright colors compared to a room where the walls are dark (makes a room safer, reduces eye fatigue when reading or working, and makes the room a more cheerful, pleasant place to be).

Overview In a house, some rooms are brighter than others, not just because they have more indoor lighting or windows to let sunlight in but because the walls, ceilings, and floors are more brightly colored. In a kitchen, people need lots of light to work with food. A bright bathroom

YOU NEED

- a room with light, brightly colored walls, which can be made completely dark
- a room with dark-colored walls, which can be made completely dark
- camera (an instant camera is preferred, but not required)
- lamp
- tape measure
- an index card or stiff piece of paper
- dark marker
- a friend

makes it a safer place. A living room, bedroom or den, however, may have darker, more deeply colored carpeting and dark walls or paneling for a quiet feeling of richness and luxury.

Bright colors, such as white and yellow, reflect much of the light that hits them. When walls, ceilings, and floors are bright in color, more light is reflected (bounced) off of those surfaces, and the light spreads around the room.

Hypothesis Photographs can be used to show how a room that has bright colored walls is brighter than a similar room that is darkly colored.

Procedure Find two rooms about the same size in a house; one room that has light-colored walls and another room that has dark walls. The rooms must be able to block any light coming in from outside of the room, such as car lights or street lights through a window, even the glare of a television set from another room. Do this experiment at night to reduce outside light from leaking in behind curtains or blinds.

Take an index card or stiff piece of white paper On one side, write "#1" with a dark marker. On the other side, write "#2."

In a room with light walls, place a lamp on a table, dresser, or any object that will raise it up to the height of a normal table. Place the table and lamp

against a wall. Turn the lamp on. Using a tape measure, stand 4 feet (122 cm) in front of the lamp, with your back to it. Have a friend stand 10 feet (305 cm) from the lamp, 6 feet (183 cm) in front of you, facing you and holding the index card with the #1 side facing the camera. You

are standing *between* your friend and the lighted lamp. Take a picture of your friend. (Do not use a camera that has an automatic flash or an automatic lens adjustment for light levels.)

Next, in a room with dark walls, place the same lamp on a table, dresser or any object that will raise it to the same height as it was in the lighter colored room. Place the table and lamp against a wall. Turn the lamp on. Again, using a tape measure, stand 4 feet in front of the lamp, facing away from it. Have your friend stand 10 feet from the lamp, face you, and hold the index card with the number #2 side facing the camera. Take a picture of your friend.

If you take or send the film for developing, tell them *not* to "adjust" prints. Also, write yourself a reminder that the #1 card photo was taken in a light-colored room and the #2 one was taken in a dark-colored room. The light source and distance of your friend from the camera remain CONSTANT. The VARIABLE is the color of the walls.

Compare the two pictures. Even though the same amount of light was used in both pictures, did the picture of your friend come out darker in the room that had the darker walls?

Results & Conclusion   Write down the results of your experiment. Come to a conclusion as to whether or not your hypothesis was correct.

Something more   Do you think the color of the ceiling and carpet on the floor also affects how light is reflected in a room?

AN UPHILL BATTLE
Kinetic energy and the transfer of energy

Purpose Demonstrate that energy can be transferred from object to object, and defy gravity.

YOU NEED
- 2 paper towel tubes
- string
- adhesive tape
- marbles
- 5 or 6 books
- modeling clay

Overview An energy force can travel like a wave, which means that it can be passed from one object to another. The force of the energy transferred can even be stronger than the force of gravity, so that the energy can be made to travel uphill, where it is possible for it to do more work.

Hypothesis Hypothesize that energy does pass through objects and this force can be transferred with enough strength to travel uphill.

Procedure By using adhesive tape at the sides, position a piece of string over the opening of one end of an empty paper towel tube. Fill the tube with marbles. Lay a thick book (such as a dictionary or encyclopedia volume) down. Tilt the open end of the paper towel tube filled with marbles up on the book. The other end of the tube, with the piece of string, should be at the bottom to keep the marbles from rolling out. Placing modeling clay on top of the book will hold the paper towel tube in place.

Stack four or more books face down on the table. Using modeling clay, tilt the other empty paper towel tube up onto the books, making a ramp. The bottom ends of both paper towel rolls must face each other, as shown in the drawings on the next page.

Roll a large marble (or two smaller ones together) down the steep paper towel tube. This rolling force is called kinetic energy, the energy of work being done. When the marble comes out of the bottom of the roll, it will hit the first marble in the next tube filled with marbles. The energy will pass from the rolling marble to the first marble, and then up through all of the

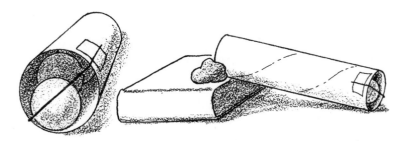

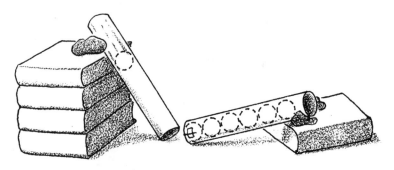

marbles. All of the marbles in the filled tube are touching each other, and while these marbles do not move, the energy passes through them. When that energy gets to the last marble at the top, the force will move the marble. Can you see it move?

Now, repeat the experiment, this time changing the slope (the angle) of the empty striking tube. Remove two of the books to lower the slope of the tube. Keep everything else CONSTANT. The only VARIABLE is the change in the slope of the striking tube. Roll the large striking marble down the tube again. Is the weaker force of the striking marble still able to move the marble at the end of the filled tube?

Results & Conclusion   Write down the results of your experiment. Come to a conclusion as to whether or not your hypothesis was correct.

Something more   Can you make the force so strong that it will knock the marble off the end? You can increase the force by using a larger striking marble (increases mass and momentum), and you can make the angle steeper for the striking marble. If you can get the last marble to roll out of the tube, how far can you get it to travel?

THE SOUND OF TIME
Amplifying sound

Purpose Since sound is sometimes hard to hear, how can it be amplified to improve audibility?

Overview Sound waves traveling through the air can be gathered to make them louder. They can also be directed (focused) in one direction to make them louder. Have you ever seen a band shell behind a large orchestra playing outside?

YOU NEED

• 2 standard-size sheets of construction paper
• adhesive tape
• a clock that "ticks" or watch with an alarm
• modeling clay
• outdoor picnic table or chair
• a friend

One way that sound is directed is by using a megaphone. A megaphone is a horn-shaped device used to increase the sound of a person's voice. Cheerleaders at a football game or the lifeguard at a beach often use megaphones. Early record players, made before the invention of electronic amplifiers, used such horns to make the music louder for listeners.

A megaphone works in reverse, too. It can gather sound and allow a person to hear weaker sounds better. Think about the shape of your outer ear, which is responsible for gathering sound. At a football game on television, you might notice a technician standing on the sidelines holding a large curved dish. This parabolic dish has a microphone attached to pick up what the players are saying.

Hypothesis Hypothesize that by using one paper megaphone at the source and another at your ear, you will be able to hear a sound that, without these devices, you would either just barely hear, or not hear it at all.

Procedure Roll a piece of paper into the shape of a horn and use adhesive tape to keep it in place. Make another horn so you have two of them.

On a quiet, calm day, go outside and find a place (table, bench, stoop) to mount your sound source. You will need a clock that has an audible "tick" or a watch with an alarm, and some modeling clay in order to stand the watch on its side.

Make sure the clock is ticking or turn the watch alarm on so that it produces its beeping sound. Move away as far as you can from the clock or watch until you can just barely hear the sound. Then stop and hold one of the horns to your ear. Can you hear the sound any better?

Now have a friend hold the other horn in front of the clock or the watch with the large end toward you. Does the beeping get even louder? The source of the sound is CONSTANT; our VARIABLES are two megaphone devices.

Note: If you do this experiment on a windy day, it could affect your results. When doing science projects, it's important to control *all* the variables, which means keep all

things constant (the same) except for those that are changed on purpose. If the wind is constant, that is, if the wind speed and direction are the same when you listen with and without the horn, maybe the results of the experiment can be trusted. But if the wind is gusting or swirling, it will very likely change the results.

Results & Conclusion   Write down the results of your experiment. Come to a conclusion as to whether or not your hypothesis was correct.

Something more   Does frequency (cycles per second) have an effect on the ability of the megaphones to amplify a sound? Use a music-instrument keyboard and compare a low note to a high note, both with and without the aid of the megaphones.

BLOWN AWAY
Fluidics: air flow around shapes

Purpose Determining how air flows around objects could sometimes be very helpful to know.

Overview It is important to understand how moving air behaves. Airplanes lift off the ground because of the way air travels past the wings, which have a special shape. When two tall buildings are close together, wind can speed up as it travels between them, causing a windy condition that may be undesirable.

YOU NEED

- handheld hair dryer
- coffee mug
- nail, about 2 inches (5 cm)
- hammer
- small block of wood
- scissors
- piece of yarn
- a pint or quart milk carton
- paper and pencil

Hypothesis The shape of an object affects how moving air flows around it.

Procedure Prove that the shape of the object affects how moving air flows around it.

Using a hammer, drive a 2-inch-long (5 cm) nail partially into a small piece of wood as shown. Near the head of the nail, tie a piece of yarn tightly onto the nail. Cut the yarn so it is about 3 inches (7 cm) long. This will be our "air-flow indicator."

Place a round coffee mug on a table. Place the air-flow indicator about 2 inches behind it.

Hold a handheld hair dryer in front of the mug and turn it on at the highest speed. Use a cool setting if it has

one. The fast-moving air splits, hugs the mug as it travels around it, and comes together behind the mug. The yarn will stand out straight like a flag or a windsock in a strong breeze, showing that air is moving quickly.

Move the air-flow indicator to various spots along the side of and behind the mug to find places where the air is moving. On a piece of paper, draw a diagram of the mug and hair dryer. Make it a view looking down from the top of the mug. Mark spots on the paper to show where there is moving air, as detected by your air-flow indicator. Use arrows to show the direction of its flow. Do you see a pattern of the air flowing around the mug?

The velocity of the moving air will be held CONSTANT. Changing the shape of the object in the stream of air flow will be the VARIABLE. What if you turn the handle of the mug to one side or the other?

Note: Since fast moving air flows around rounded objects and meets behind it, maybe hiding behind a tree or telephone pole to block wind is not as effective as you might think!

Now, replace the mug with a rectangular-shaped object, such as a pint or quart carton of milk. Again, move the yarn air-flow indicator around the carton and draw a diagram showing the air flow around it.

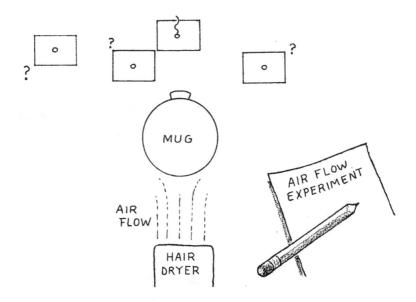

Results & Conclusion Write down the results of your experiment. Come to a conclusion as to whether or not your hypothesis was correct.

Something more Making the piece of yarn longer will allow it to become an indicator of air speed. The farther out the yarn is blown, the stronger the air flow. Now you can use the device to compare the strength of air flow in addition to direction.

BALANCING ACT
Objects at rest tend to stay at rest

Purpose We want to find out if objects at rest will remain at rest if balanced forces are applied.

Overview Sir Isaac Newton discovered some important principles of natural science. One of the laws of nature he recognized is that "objects at rest tend to stay at rest," which means that things that are not moving will stay that way unless an "unbalanced force" pushes (or pulls) on them. An unbalanced force is a push on an object that is stronger in one direction than a push from the opposite direction.

Balanced forces occur when equal forces coming from opposite directions are applied to an object. A

YOU NEED

- an adult
- a 9-inch (21 cm) wooden dowel, an inch or more (3 cm) in diameter
- hand wood saw
- 3 small nails
- hammer
- thin string
- scissors
- broom handle or yardstick
- 2 chairs of equal height

book lying on a table has balanced forces acting upon it; the table is pushing up and gravity is pulling down. If the forces are equal, but they are not at exactly opposite angles, the resulting force will be unbalanced. In other words, the book may slide down the slanted table.

Hypothesis Objects at rest tend to stay at rest when balanced forces are applied to them.

Procedure To put together a device to prove, or disprove, the above hypothesis, ask an adult to help by cutting a wooden dowel into three 3-inch-long (7 cm) cylinders. The dowel should be at least 1 to 1½ inches in diameter. If the dowel is purchased at a hardware store or hobby shop, the sales clerk may offer to cut the dowel to size for you.

   With a hammer, tap a small nail into the top of each dowel at exactly the middle. To find the middle accurately, you can use two pieces of string or thread. Lay one piece across the end of the dowel and another straight across it, at a 90-degree angle. The point where the two pieces cross each other is the middle.

Tie one end of a 3-foot-long (90 cm) piece of string onto the nail of one of the dowels. Do the same for the other two dowels.

Place two chairs equal in height back to back (such as matching kitchen or dining room chairs), but separate them by about 3 feet. Lay a broom handle across the top of the chair backs (a yardstick or any long, stiff pole will also work).

Tie the loose end of one of the dowels to the center of the broom handle, so that the dowel hangs down, but does not touch the floor.

Hang another dowel on the right and another one on the left of the first dowel, so they are side by side.

Tie the ends to the broom handle so that all three dowels are hanging straight and just touching alongside each other when they are not moving.

Take the left dowel in your left hand and the right dowel in your right hand. Pull them both away from the center hanging dowel until they are each about a foot (30 cm) away from the center dowel. Let the dowels in each hand go at exactly the same moment (this may take some practice), so they will both hit the center dowel together.

If only one dowel should swing into the center dowel (which is at rest), the "unbalanced force" will push the center dowel and make it swing, too. But, if both swinging dowels apply *equal* force in *opposite* directions, the experiment should result in a balanced force on the center dowel so that it remains at rest.

Results & Conclusion Write down the results of your experiment. Come to a conclusion as to whether or not your hypothesis was correct.

Something more Show balanced and unbalanced forces in the game of "tug of war." Tie a ribbon onto the center of a long rope. Place a brick or some object on the ground to mark a spot, and lay the rope over it, with the ribbon on top of the brick. Have several friends grab one end of the rope and several grab the other for a game of tug of war. When equal pulling force is on both sides, the ribbon will stay hovered over the brick. When the pulling force becomes unbalanced, the ribbon will move toward the friends who are exerting the stronger total force.

FLOATING ALONG
Buoyancy: the ability to stay up

Purpose Let's figure out how some things that are heavier than water can float.

Overview Why do things float? An object may float in water if it is light and weighs less than water. A Ping-Pong ball will float because it weighs less than water.

But why does a heavy boat float? Big ships are made out of steel, and steel is much heavier than water.

The "buoyancy" of an object is its ability to float on

YOU NEED
- modeling clay
- large bowl
- water
- small kitchen food scale (gram weight scale)
- small bowl or cup
- kitchen measuring cup that has a pour spout
- thin piece of thread

the surface of water (or any fluid). Water gives an upward push on any object in it. The amount of force pushing upward is equal to the weight of the water that the object "displaces" (takes the place of).

So, if a boat or ship is designed to displace an amount of water that weighs more than the boat, it will be able to float. For something to be buoyant, its shape is very important.

Hypothesis An object that is heavier than water can be made to float.

Procedure Fill a large bowl with water, but don't fill it all the way to the top. Take a small amount of modeling clay and use your hands to roll it into a ball that is about two inches (5 cm) in diameter. Place the

ball on the surface of the water and let go. The clay ball is heavier than water. Does it float or sink?

Take the ball out of the water. Use your hands to mold the same clay into the shape of a small boat. It should have a flat bottom and sides. Now, place the boat on the surface of the water. It floats, even though it is the same amount of clay. While holding the weight (mass) of the clay CONSTANT, the VARIABLE has been its change in shape.

You can take this project further by capturing and weighing the water that is displaced by the ball of clay. To do this you will need a kitchen measuring cup that has a pour spout. Set a small bowl or cup under the spout to catch the water that spills out.

Fill the measuring cup with water until water begins to spill out of the spout. When the water stops overflowing, empty the small bowl that caught the water. Dry it out.

Shape the clay into a ball. Tie a piece of thin thread onto the ball and slowly lower it into the water. The water it displaces will spill out into the bowl.

When the water stops overflowing, remove the bowl and weigh it on a small kitchen food scale. Write down the weight.

Dry the bowl and weigh it. This is to get the "tare weight," the weight of the container that had been holding the water. Subtract this tare weight from the weight of the bowl with the water in it. The difference is the weight of the displaced water.

Also weigh the clay ball. Compare the weight of the clay ball to the weight of the water that it displaced.

Results & Conclusion Write down the results of your experiment. Come to a conclusion as to whether or not your hypothesis was correct.

Something more Weigh the amount of water the clay boat displaces. Compare it to the amount of water the same clay in the shape of a ball displaces. Do you think the weight of the water displaced by the boat will be less than the weight of the water displaced by the ball?

UNWELCOME GUSTS
Comparing and measuring wind strengths

Purpose Over a week's time, we want to determine which day had the strongest gust of wind.

Overview People have harnessed the powerful force of wind to help them do many useful things. Windmills have been used to pump water and make electricity. Boats can sail around the world by using sails whose large surface area captures the wind's power.

But, sometimes this wind force works against us. On extremely windy and gusty days, huge bridges are sometimes closed to vehicles with large surface areas, such as tractor trailers and motor homes, because of the danger. Playing beachball or volleyball in a strong wind can either help your

YOU NEED

- 7 medium-size (8 oz.) plastic or foam cups
- piece of board
- 2 cinder blocks/trash cans
- a wide-open area, away from buildings
- water
- kitchen measuring cup
- heavy mug or old pot

team or hinder it, depending on whether or not you are downwind.

On a windy day, have you ever tried to carry a big piece of plywood or poster board, or a small bag holding only something light, like a greeting card? Have you ever helped your parents put garbage cans out at the curb on a windy trash day? If no one is home when the trash is collected, you could come home to find the cans scattered all over. The empty cans, being lighter without trash, are easily blown all over the yard and even into the street, causing a hazard to motorists, by strong gusts of wind. Can you think of other times when the force of the wind is not welcome, or is even harmful?

Wind velocity is important in air travel. In physics, the word "velocity" means both speed and direction. Airports use wind socks to get a relative indication of wind direction and speed.

You can build a simple weather instrument to detect wind gusts and get a comparative indication of their strength by using water-filled paper cups.

Hypothesis Hypothesize that you can put together a simple device that will allow you to determine the force of the strongest wind gust of the day.

Procedure Find an open area away from buildings or other structures that might block the wind. A spot in your own backyard would be good, if space is available.

Set two cinder blocks upright on the ground several feet apart. If you don't have cinder blocks, you can use same-size milk crates, or buckets or trash cans turned upside down. Across the top of any such "risers," lay a long piece of wood.

Set seven 8-ounce paper cups (or plastic cups) in a row on the board. Leave one empty and, using a

measuring cup, pour 1 ounce of water in the second cup, 2 ounces in the next, 3 in the next, continuing up to 7 ounces. If your measuring cup is marked in milliliters, use increments of 50 (that is, 50 ml, 100 ml, 150 ml, 200 ml, and so on).

On the board or in an open place nearby, place a heavy mug or old pot. This will be used to capture any rainfall for the day. If you find rain in it, don't record that day's results because they won't be valid.

At the end of each day, observe which cups have blown off the board. Write down how many cups blew off. The lighter cups are more sensitive to the force of the wind.

The next day, set them up again, and refill the cups with water (some water will probably have evaporated). The contents and positions of all of the cups must be kept *constant*. The wind gusts should be the only *variable* in the project. At the end of the day, record which cups have blown off the board.

Do this every day for a week or two, or as long as you wish. Look at your recorded observations for each day. Did your paper-cup system work as a weather instrument, to allow comparisons of the strongest gusts of wind that occurred each day?

Results & Conclusion Write down the results of your experiment. Come to a conclusion as to whether or not your hypothesis was correct.

Something more What if you found a heavy and a light cup knocked over, but one in the middle still standing? Do you think the results of that day should not be used, as there may have been interference from squirrels, birds, or other animals seeking water?

BREAK THE BEAM
Exploring some characteristics of light

Purpose The purpose is to understand how a light security system works.

Overview Can you see a beam of light? You can certainly see a lit light bulb, the sun, a candle's flame, or any source of light. You can also see objects because light is shining on them. But, you are normally not able to see the light beam itself. The path of the beam through the air can, however, be seen by filling the air with tiny particles so the light will reflect off them.

YOU NEED
- lamp
- handheld mirror
- modeling clay
- small piece of cardboard
- flashlight
- several facial tissues
- dark room
- a friend
- table

In a dark room, lay a flashlight on a table and shake several facial tissues in the air. The light will reflect (bounce) off of the tiny particles of tissue, allowing you to see the path of the light beam. Similarly, have

you ever seen the rays of sunlight shine from behind
breaks in clouds? Have you ever seen the light beams
coming from the headlights of your car on a dark,
foggy evening?

Because we can't actually see a beam of light, some
home security systems use light to detect if someone
walks through a room. A light source shines into an
electronic device that detects the light. Everything is
fine as long as light is shining on it. But the light beam
is interrupted when someone walks between the light
source and the security device, it senses that the light
has gone out, and it sounds an alarm. To make the

light less detectable by anyone, a red filter is used to reduce the light reflected by any tiny particles that may be in the air.

Hypothesis   We can detect a person walking in another room using a light source and a mirror.

Procedure   Set up a demonstration of how a home security system might work. Lay a hand mirror on its side on a piece of cardboard. With modeling clay, build a base around it so the mirror will stand up by itself, as shown.

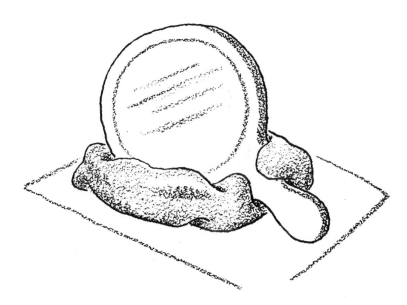

Set the mirror on a table or bookshelf, and adjust it so that the light from a lamp is reflected into another room. It should be a room that you can make fairly dark. Stand against the wall and look out the door at the mirror. Have a friend adjust the mirror until you can see the lamp in it. Then you know the mirror is lined up. This demonstrates another characteristic about light; light travels in a straight line unless something interferes.

Go into the dark room and close the door until it is only open enough to let a slit of light in to shine against the wall opposite the door.

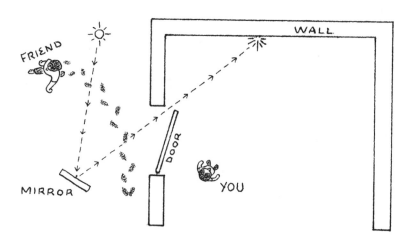

Watch the light on the wall as you have your friend walk around the room. Can you detect when your friend steps in the path of the light?

Results & Conclusion   Write down the results of your experiment. Come to a conclusion as to whether or not your hypothesis was correct.

Something more   You can tell the direction of a person walking through the room by adding a second mirror a few feet to either side of the first mirror, so that a second light spot shines on the wall. Then, if a person walks by, they will break one beam before the other. The beam that is broken first tells you in which direction the person is walking.

CRASH!

The relationship between mass and force

Purpose It often happens that objects that are at rest, that is, not moving, are hit by moving objects and forced to move. What happens when the objects struck have different masses?

Overview Sir Isaac Newton did experiments to find the mathematical relationship between the mass of an object and how fast it moves when a given force strikes it. Mass is how much "stuff" an object is made up of. Newton found that the larger the mass of an object, the smaller will be its movement when a given force is applied.

> **YOU NEED**
> - an adult
> - ladder
> - soccer ball
> - bowling ball
> - golf ball
> - several thick hardback books
> - 2 wooden boards, 2-by-4-inch by 8 feet (240 cm) long
> - hammer
> - 6 long nails
> - ruler

Imagine a soccer ball filled with air and another one that is filled with sand. If you kicked each soccer ball with the same amount of force, the ball with more mass (filled with sand) would not move as far as the one with less mass (filled with air). You might hurt your foot on the ball with more mass, too!

Hypothesis A soccer ball will move farther than a bowling ball when the same force is applied to each.

Procedure Let's strike a bowling ball (in place of a sand-filled soccer ball) and a soccer ball with the same force and measure how far they each move. If Sir Isaac Newton is right, the soccer ball, which has less mass than the bowling ball, will move farther. (Be very careful handling the bowling ball. It could hurt your foot if it should fall on it. Have an adult help you if the ball is too heavy for you to handle safely.)

We need to have a force that will be exactly the same every time, so we can be sure each ball is struck with an identical force. This is our CONSTANT. To do this, construct a ramp with the long two 2-by-4-inch wooden boards, making a "V" shape. Rolling a golf ball down the "V" channel will cause it to strike whatever object is at the bottom of the ramp with the same force every time. If we let go of the golf ball at the same place on the ramp each time, the force of gravity will ensure that the ball is rolling at the same speed every time it reaches the bottom of the ramp.

Nail two 2-by-4, 8-foot-long pieces of lumber together, making a "V" shaped channel. This will be our wooden ramp.

Outside, set up a ladder. Rest one end of the ramp on the third or fourth rung of the ladder. At the ground end of the ramp, place a bowling ball so that it is touching the end of the ramp. The ground must be flat and level.

You'll get the most action if the bowling ball is struck in its middle, some small distance above the ground. Place books underneath

the end of the ramp to raise it until it is positioned at
the middle of the bowling ball. You may need to place
a few books along the sides of the ramp to keep the
"V" shape facing up.

Pick a spot along the ramp to let go of a golf ball and
start it rolling down the ramp. To get the most speed
out of the ball, you can let it go from the high end of
the ramp. Be sure, however, that you let the ball go
from the same spot every time. Also, don't give the golf
ball a "push" start, because you would not give it an
even push every time. Just let go of the ball and gravity
will start it rolling.

If the bowling ball moves when it is hit, use a ruler
to measure how far it moved.

Now we want to see how far a soccer ball, which
has much less mass, will roll when the same force is
applied to it. The different masses of the two balls is
the VARIABLE in our project.

Release the golf ball. If the soccer ball moves a lot,
it may be easier to use a tape measure, yardstick, or
meter stick than a ruler to measure the distance it
rolled.

If neither ball moved, increase the slope of the ramp by moving up one rung on the ladder, giving more speed, hence force, to the rolling ball.

Results & Conclusion   Write down the results of your experiment. Come to a conclusion as to whether or not your hypothesis was correct.

Something more   Repeat this experiment using different balls at the bottom of the ramp; try a baseball, a basketball and a tennis ball. Can you predict which one will roll farthest?

BIGGER WATER

Temperature's expansion/contraction effects

Purpose What happens when water freezes?

Overview Many things expand or contract when they change temperature. Have you ever noticed, when standing by a railroad track, why there are

gaps in the rails at certain intervals? The spaces in the rails allow them room to "grow," in case they expand; otherwise, the rails would buckle. Track engineers know exactly how big the gaps should be to allow for this rail expansion.

If you have electric baseboard heat in your home, you may have heard the crackling sounds it makes when the metal fins heat up or cool down. That's because the fins are expanding and contracting. Gaps in the roadway of bridges are also there to allow for expansion and contraction from temperature changes.

Hypothesis The same amount of water takes up more space when it is frozen.

Procedure Fill an empty soup can with water; be careful of sharp can edges. Set the can in a small bowl, and place it in the freezer section of a refrigerator. The bowl will catch any water that might spill from the can. Add more water if necessary so the water level in the can is at the very top. Leave the can of water in the freezer overnight.

In this experiment, the quantity of water is being held CONSTANT, and the temperature is our VARIABLE.

Take the can of ice out of the freezer in the morning. The volume of water that fit into the can when it was a liquid is now too big for the can. The ice has risen above the top of the can because of the expanding water and its push against the bottom of the can.

Results & Conclusion Write down the results of your experiment. Come to a conclusion as to whether or not your hypothesis was correct.

Something more Quantify how much more volume the ice takes up than it did as a liquid by using displacement, which is explained in the project "Floating Along." Hold the can tightly to melt the ice slightly around the edges of the can so the ice will come out as one block. Dip the block in a container of water and measure the water that it displaces.

WORK = FORCE x DISTANCE
The wedge, a simple machine

Purpose Show the relationship between the distance a wedge is moved forward and the height an object sitting on top of the wedge is raised.

Overview A "wedge" is one of those "simple machines" we talked about. A wedge is an object in the shape of a triangle. A doorstop and the metal head of an axe are examples of wedges.

When an axe or chopping maul is used to split firewood, the worker swings the tool over a large distance to strike the wood with great force. That force is turned into the small distance covered by the wedge, as the axe moves down into the wood to split it.

YOU NEED
- a wide strip of thick cardboard, about 12 inches (30 cm) long
- two rulers
- small, light cardboard box (shoebox or a similar size)
- scissors
- pencil
- paper
- 1 or more heavy books

In science, "work" is a measurement equal to "force" times "distance."

When a force is applied to a wedge, the force moves the wedge forward, but it also moves anything resting on top of the wedge into an upward direction (at a 90-degree angle to the forward movement of the wedge).

A wedge can be used to lift very heavy objects a short distance. House movers sometimes use wedges between the sill plate and the foundation to raise a house up so steel girders can be slid under it.

Hypothesis Using a wedge increases the amount of force in a perpendicular direction, but we pay for it in a decrease in distance.

Procedure Draw a right triangle on a thick piece of cardboard. Make the triangle about 2 inches (5 cm) tall by about 12 inches (30 cm) in length. The hypotenuse of the triangle will form a long, gently sloping ramp. Use scissors to cut the triangle out.

Place a small, light box on a table. A shoebox would be perfect. At one end of the box, stack one or two heavy books. That will keep the box from sliding.

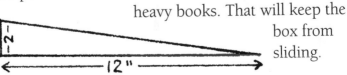

At the other end, place your cutout wedge so that its pointed tip just slips under the box. Lay a ruler alongside the box, with the zero mark on the ruler at the edge of the box where the wedge touches the box. The length of the ruler should face away from the wedge (running parallel to the side of the box).

Push on the wedge so that it slides 2 inches (5 cm) under the box. Use another ruler to measure how high the end of the box is raised above the table.

On a piece of paper, draw two vertical columns. Label the heading on one column DISTANCE WEDGE MOVED and the other column HEIGHT RAISED. Write the measurement under the first heading and the distance the box was raised in the second column.

The CONSTANT in this project is the incline (the slope) of the wedge, the box it is lifting, and the force applied. The VARIABLE

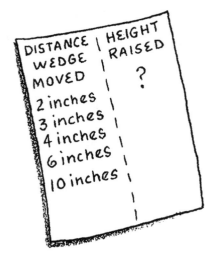

DISTANCE WEDGE MOVED	HEIGHT RAISED
2 inches	?
3 inches	
4 inches	
6 inches	
10 inches	

is the distance the wedge is moved inward and the height it pushes up on the box.

Now push the box forward another inch or centimeter, and record the height raised. Continue to push the wedge under the box at each increment, until the top of the wedge is reached. Write down the distance and the height for each move increment.

Results & Conclusion Write down the results of your experiment. Come to a conclusion as to whether or not your hypothesis was correct.

Something more A wedge doorstop is a stationary wedge that is applying a force equal to the force needed to keep the door from closing.

A nail is also a wedge. Can you imagine pushing something into a piece of wood that doesn't come to a point? The smaller the nail, the easier it is to wedge

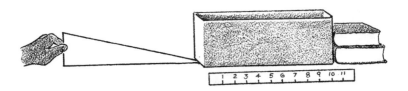

into the wood, because it has less wood material to push out of the way. Try pushing a nail into a piece of wood by hand. Then try pushing a thumbtack with a head on it into the same piece of wood. Is the thumbtack much easier to push in?

WATER SHOOTER
Compressing a fluid

Purpose Let's learn a little about "hydraulics."

Overview Hydraulics, from the Greek word meaning "about water," is the study of liquid in motion. One element of hydraulics deals with compressing a liquid, which is used in machines where great pushing or lifting strength is needed. A force pushing on any part of an enclosed liquid creates an equal pressure per unit of area on everything the liquid touches. By using a system of pistons (cylindrical containers filled with a liquid), great force can be achieved.

Hypothesis When water is compressed and forced to flow out of an opening, the velocity of the water will be much greater if the opening is small than if it is large.

YOU NEED

- squeeze bottle with spout top
- a piece of heavy board
- water
- ruler

Procedure Outside, remove the spout and fill the
squeeze bottle with water. Set the board on its side and
tilt the container against it. Let water leak out until it
stops; the angle will keep most of the water inside.
Then with your hand, strike the side of the bottle,
forcing water out through the opening. Watch how far
the stream of water shoots.

Again, fill the squeeze bottle with water, set it
against the board and let the water leak out. This time,
screw on the spout. Keeping the volume of water
CONSTANT (by tilting) as well as the striking force used
on the bottle, our VARIABLE will be the diameter of the
opening through which the water escapes. The spout
makes the opening much smaller. Strike the side of the
container again, using the same amount of force as

before. Does the stream of water travel farther with the spout on? Does that mean the velocity (speed) of the water coming out was greater?

Results & Conclusion Write down the results of your experiment. Come to a conclusion as to whether or not your hypothesis was correct.

Something more Compare the amount of water coming out of the bottle with the spout off and with it on. Use a measuring cup to quantify the volume of water in the bottle before and after each strike. Perhaps, with the spout on, less water is coming out? Did you strike the bottle with the same force each time? Can you find a way to be sure?

SIPHON FUN
Water drains to its own level

Purpose Discover how a siphon works.

Overview A hose or tube can be used to create a "siphon," a device that drains liquids. A siphon will drain liquid from a higher to a lower level, even if it first has to travel uphill! A hose placed with one end underwater in an above-ground swimming pool and the rest draped over the side and down to the ground will drain water out of the pool. To get the flow started, it may be necessary to suck on the low end of the hose. Once the flow begins, gravity pulls the liquid down, creating a vacuum in the tube that draws the liquid up and through the tube.

Hypothesis Water can be made to rise above its level through the use of a siphon.

Procedure Stack books on a table. Fill a clear plastic bottle with water and place it on the books. Set an

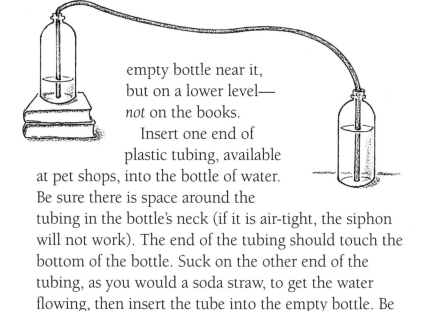

empty bottle near it,
but on a lower level—
not on the books.

Insert one end of
plastic tubing, available
at pet shops, into the bottle of water.
Be sure there is space around the
tubing in the bottle's neck (if it is air-tight, the siphon
will not work). The end of the tubing should touch the
bottom of the bottle. Suck on the other end of the
tubing, as you would a soda straw, to get the water
flowing, then insert the tube into the empty bottle. Be
careful with the tubing in your mouth; remove before
water reaches the end of tube you had sucked on. Push
the tubing in so it touches the bottom of the bottle.

Held CONSTANT is gravity, the volume of water, the
lack of air in the tube system, and the bottles. The
motion of the water from higher to lower bottle is the
VARIABLE.

Results & Conclusion   Write down the results of your
experiment. Come to a conclusion as to whether or not
your hypothesis was correct.

Something more

1. Once the lower bottle is filled, do you think that the
   siphon would work in reverse? What if the full
   bottle was now raised higher than the one on the
   books?

2. Does it take the same amount of time for the bottle
   to empty each time the experiment is done, so that it
   could be used as a sort of "water clock"?

BACK IN SHAPE
The characteristics of elasticity

Purpose To find the effect of weathering and stretching on the elasticity of a rubber band.

Overview One characteristic of a material is a measure of its "elasticity." Elasticity is the ability of a material to return to its original shape after it has been stretched or pressed together. Balloons and rubber bands are very elastic. Can you think of other examples?

YOU NEED

- 2 identical rubber bands
- 2 paper clips
- large empty milk carton
- water
- clothesline
- nail or awl, to poke a small hole in the milk carton

Have you ever seen a rubber band wrapped around a newspaper or something else that has been outside for some time? Have you ever seen an old rubber band that has been wrapped around something in an attic for many years? Can you observe cracks in the bands? Are they able to be stretched? Do they snap back to a smaller shape?

Hypothesis After a rubber band has been stretched and exposed to outdoor weathering elements for several days, the elasticity of the rubber band will be affected, and it will not return to its original shape.

Procedure For this project, we'll need a device to test the elasticity of the material.

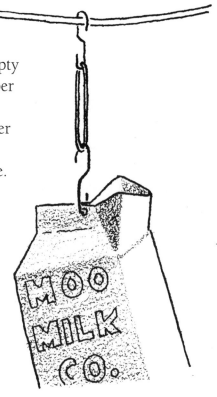

Gather two identical rubber bands, a large empty milk carton, and two paper clips. Lay the two rubber bands on top of each other to check that they are as equal in shape as possible. The rubber band will be our CONSTANT. Our VARIABLE will be to stretch one rubber band and expose it to weather, while the other will remain unstretched and indoors, protected from weathering.

Bend open the paper clips to make "S" shaped hooks. Carefully poke a hole in the top of a large empty milk carton, and insert a paper clip hook through it.

Hang another paper clip hook on an outdoor clothesline and drape a rubber band on the bottom part of the paper clip hook. Hang the milk carton on the rubber band using the paper clip hook in the carton, as shown on the next page.

Pour some water into the milk carton. This will create a weight to stretch the rubber band. We want to stretch the rubber band a lot, but not to the breaking point. Since rubber bands are not all the same, we can't tell you how much water to add to the carton. (Science is not always like a food recipe!) Slowly add more water to the milk carton until it looks like the rubber band is well stretched, but not in danger of breaking.

Leave the rubber band stretching device hanging on the clothes line for three or four days. Then, carefully, take the device apart and remove the rubber band. Line it up next to the other identical one that you had put aside indoors and compare them. Has the stretched one changed shape? Is it able to completely return to its original shape?

Inspect the stretched rubber band closely. Are you able to see any cracks, discolorations, or other signs of deterioration caused by the experiment?

Results & Conclusion Write down the results of your experiment. Come to a conclusion as to whether or not your hypothesis was correct.

Something more Can you use your stretching device to test the elasticity of other materials or objects?

What happens to a rubber band stretched as in this experiment, but indoors?

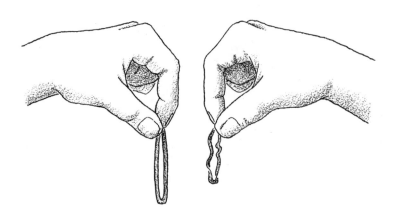

DOWN-RANGE SHOOTER
Trajectory: curved path through the air

Purpose The purpose of this experiment is to determine the angle of trajectory that will give the greatest distance.

Overview When you throw a stone a little upward and away from you, you know that it will not keep going in that direction, but will slowly curve and begin to fall to the ground. The path of an object hurled through the air is called its trajectory. This path is caused by the motions of the stone, moving forward and moving upward at first, then downward. The stone moves upward and forward because of the force of your throw. But, because the stone has weight, the Earth's gravity eventually

causes it to curve and fall down.

What determines where an object will land when it is thrown or launched? The force at which it is thrown and its upward angle are both factors.

The trajectory of an object is very important to the military when they use artillery. If a cannon is to hit its target, the operator has to know the right angle to tilt the gun upward.

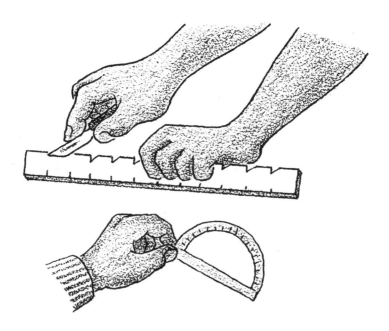

Hypothesis Hypothesize that the launch angle of an object will affect the distance it will travel away from the launching device.

Procedure Because of the hazard of using a sharp tool, have an adult cut a small notch at each inch or centimeter mark on a wooden ruler with a utility knife or razor.

Stretch a rubber band from one end of the ruler to one of the notches at a marking, giving it a good stretch, but not to its maximum stretch potential. Hold the ruler lengthwise with one hand near the edge of a table. With the other hand, push the rubber band out of the notch until it launches. Place a domino on the floor to mark the spot where it landed. Always keep safety in mind; do not launch the rubber band while anyone is standing in front of the ruler.

Then raise the end of the ruler by placing dominos under it. Experiment with different heights (elevation). By launching the rubber band from the same marking, the launch force is kept CONSTANT, and only the angle is VARIABLE.

Set a protractor on the table and measure the angle the ruler is elevated upward before each launch. What

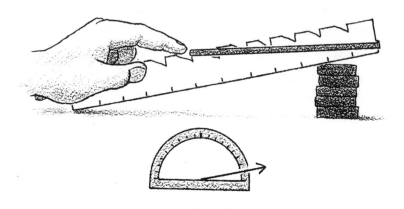

is the angle that shoots the rubber band the farthest distance from the ruler? What is the angle at which, if the angle is further increased, the rubber band will not travel any farther (this is the optimum angle for distance)?

Note: If the rubber band gets caught on some of the other notches in the ruler as it launches, you can put tape over them, or make a horizontal cut in a straw and cover the notches.

Results & Conclusion   Write down the results of your experiment. Come to a conclusion as to whether or not your hypothesis was correct.

Something more

1. The distance the rubber band will travel depends on both the launch angle and the launch force. Try launching the rubber band from different notches, which changes the force. The more the rubber band is stretched, the more force it will have when it is launched. What combination of tilt angle and launch force (marked notch) makes the rubber band travel the farthest? The highest?

2. Make a game by having a friend place a domino on the floor while you try different combinations of force and tilt to land the rubber band as close as you can to the target. Then reverse places with your friend, and see who can come closest with the least number of tries.

INDEX

Illustration Credits

Illustrations appearing in this book between pages 12–97 and 196–280 are by Frances Zweifel

Illustrations appearing in this book on pages 105–189 are by Alex Pang

WHAT IS MENSA?

Mensa—The High IQ Society

Mensa is the international society for people with a high IQ. We have more than 100,000 members in over 40 countries worldwide.
The society's aims are:

- To identify and foster human intelligence for the benefit of humanity;
- To encourage research in the nature, characteristics, and uses of intelligence;
- To provide a stimulating intellectual and social environment for its members.

Anyone with an IQ score in the top two percent of the population is eligible to become a member of Mensa—are you the "one in 50" we've been looking for?
Mensa membership offers an excellent range of benefits:

- Networking and social activities nationally and around the world;
- Special Interest Groups (hundreds of chances to pursue your hobbies and interests—from art to zoology!);
- Monthly International Journal, national magazines, and regional newsletters;
- Local meetings—from game challenges to food and drink;
- National and international weekend gatherings and conferences;
- Intellectually stimulating lectures and seminars;
- Access to the worldwide SIGHT network for travelers and hosts.

**For more information about
Mensa International:**

www.mensa.org
Mensa International
15 The Ivories
6–8 Northampton Street
Islington, London N1 2HY
United Kingdom

**For more information about
American Mensa:**

www.us.mensa.org
Telephone: (800) 66-MENSA
American Mensa Ltd.
1229 Corporate Drive West
Arlington, TX 76006-6103 US

**For more information about
British Mensa (UK and Ireland):**

www.mensa.org.uk
Telephone: +44 (0) 1902 772771
E-mail: enquiries@mensa.org.uk
British Mensa Ltd.
St. John's House
St. John's Square
Wolverhampton WV2 4AH
United Kingdom